Unmanned System Technologies

Springer's Unmanned Systems Technologies (UST) book series publishes the latest developments in unmanned vehicles and platforms in a timely manner, with the highest of quality, and written and edited by leaders in the field. The aim is to provide an effective platform to global researchers in the field to exchange their research findings and ideas. The series covers all the main branches of unmanned systems and technologies, both theoretical and applied, including but not limited to:

- Unmanned aerial vehicles, unmanned ground vehicles and unmanned ships, and all unmanned systems related research in:
- Robotics Design
- Artificial Intelligence
- Guidance, Navigation and Control
- Signal Processing
- Circuit and Systems
- Mechatronics
- Big Data
- Intelligent Computing and Communication
- Advanced Materials and Engineering

The publication types of the series are monographs, professional books, graduate textbooks, and edited volumes.

More information about this series at https://link.springer.com/bookseries/15608

David Sacharny • Thomas C. Henderson

Lane-Based Unmanned Aircraft Systems Traffic Management

 Springer

David Sacharny
University of Utah
Salt Lake City, UT, USA

Thomas C. Henderson
University of Utah
Salt Lake City, UT, USA

ISSN 2523-3734 ISSN 2523-3742 (electronic)
Unmanned System Technologies
ISBN 978-3-030-98573-8 ISBN 978-3-030-98574-5 (eBook)
https://doi.org/10.1007/978-3-030-98574-5

This Springer imprint is published by the registered company Springer Nature Switzerland AG
The registered company address is: Gewerbestrasse 11, 6330 Cham, Switzerland

To all those who participated in developing the ideas and systems presented here (especially to Vista Marston), to our families, and to the future of autonomous systems!

Preface

This book is the result of many years of effort in trying to develop an efficient and effective approach for large-scale UAS traffic management. The methods we present apply to a future of air mobility that imagines a dense network of autonomous aircraft, transporting people and things within and between cities. Throughout the book, we make connections to the ground transportation network, and we take inspiration from the engineering that has developed there over the last century. We combine aspects of ground traffic engineering with the latest research in advanced air mobility.

At the time of writing of this book, advanced air mobility is still in its infancy. This is apparent by the absence of low-altitude vehicles flying overhead, but also by the lack of standardization and the pervasive questioning of whether such a future is yet possible. Among all the problems that motivate this skepticism, the problem of automating air traffic control is particularly interesting to the authors of this book. It is an engineering problem that is susceptible to catastrophic consequences due to computational intractability, and so demands the attention of researchers in computer science and robotics. The current iteration of the autonomous air traffic control system proposed by NASA and the FAA draws heavily from current human air traffic control practices.

Both ground transportation and air traffic control systems incorporate many trade-offs when it comes to safety, reliability, and innovation. However, they share the characteristic of relying on human cognition to make critical decisions. If advanced air mobility requires aircraft to fly autonomously, then it follows that those critical decisions must be made predictably by machines. Just as roundabouts can replace signalized intersections, thereby reducing the coordination complexity for humans, the lane-based approach is an attempt to simplify the environment for autonomous vehicles.

Much remains to be done, and we have tried to point out research directions at the end of each chapter. Thus, this book should provide some guideposts to the future of UAS traffic management as well as an exposition of the current state of the art. We look forward to participating in discovering that future!

Salt Lake City, UT, USA   David Sacharny
Salt Lake City, UT, USA   Thomas C. Henderson

Contents

Chapter 1
Current State of Affairs: Economic Impact

1.1 Motivation

Our research team entered the Advanced Air Mobility (AAM) arena in August of 2018, when Andrew W. Buffmire, Research Corporate Ambassador for the College of Engineering at University of Utah, invited our research team to a meeting titled "UAS Modeling and Management," held in the Halverson Conference Room in the Warnock Engineering Building. Among the twelve attendees was Jared Esselman, the Director of Aeronautics at the Utah Department of Transportation (UDOT), as well as a representative from the Governor's Office of Economic Opportunity, the University of Utah Hospital System, the university's Entertainment Arts Engineering program, and Fortem Technologies. Before the meeting, an invitee from Utah's Automated Geographic Reference Center (AGRC) sent an email to the group apologizing in advance and providing some data that foreshadowed the discussion:

```
Hello all,

I am unable to attend the meeting this Friday, but
wanted to send out a few maps that may be helpful while
having this discussion and let you know that we have
access to several GIS datasets that may be helpful in
putting together the air space model.

Please let me know how I can help.

Sean
```

© The Author(s), under exclusive license to Springer Nature Switzerland AG 2022
D. Sacharny, T. C. Henderson, *Lane-Based Unmanned Aircraft Systems Traffic Management*, Unmanned System Technologies,
https://doi.org/10.1007/978-3-030-98574-5_1

Sean Fernandez, PLS
State Cadastral Surveyor/Division Manager
State of Utah AGRC/DTS

In one attachment (Fig. 1.1), a map of features and the Salt Lake valley included locations of GPS sensors for Real-Time Kinematic (RTK) positioning, liquor stores, libraries, post offices, correctional facilities, and schools. At the meeting, we discussed how the GPS network could be used by autonomous vehicles to accurately report their telemetry and maintain safe separation. The other features on the map were endpoints in an imagined airspace network that transported cargo and passengers quickly, quietly, and efficiently between locations within city limits. We treated the availability of vehicles that could meet the challenges of urban air mobility (UAM) as an inevitability, and we focused on how our city could become a platform for developing this new transportation system. In Utah, a well-established and growing aeronautics and technology industry, research institutions with a running start on robotics, and a forward-looking governmental body at UDOT provided a means for making serious contributions to this vision of the future.

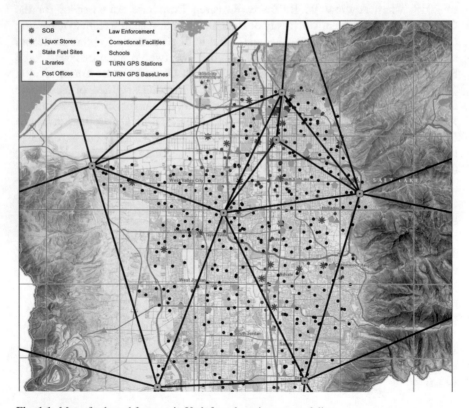

Fig. 1.1 Map of selected features in Utah for urban airspace modeling

However, it was also clear that this vision entailed considerable technical, logistical, and business challenges to overcome.

Advanced Air Mobility describes an emerging aviation market for local, regional, intraregional, and urban use-cases, supported by a set of disruptive technologies. The most salient technologies are the vehicles, and the promise that they will fly themselves effortlessly throughout the city has generated billions of dollars in private and public investment. Just a few of the publicly traded companies in this space, Joby Aviation, Lilium, Archer, and Volocopter represent over $4 Billion in market cap and have yet to transport a single paying customer. With so much investment and engineering effort going into vehicle development, it is no wonder that early adopters and innovators are bullish, and it is easy to imagine, given the state of traffic in many cities, that a scenic ride inside a Tesla-like aircraft would be popular (assuming the price was right). Within each vehicle also exist numerous other disruptive technologies, electric and hybrid propulsion systems, energy storage systems, guidance and control software, advanced materials, etc. Each of these systems must interoperate or contend with an ecosystem of other vehicles and disruptive technologies in infrastructure, simulation, monitoring, and air traffic management. The minimum set of disruptive technologies necessary to enable this vision of urban air mobility is a subject of debate, and in the United States, it will be determined by the businesses that are commercially successful. Disruptive technologies are innovations that alter the way people and industries operate, and the technologies that transform urban mobility are as certain to be disruptive as when selective availability was discontinued for GPS in the year 2000. Unlike GPS, however, the trajectory to enable mass adoption and commercial viability is much less clear.

In 2018, NASA hired two companies, Crown Consulting, Inc., and Booz Allen Hamilton, to study the market viability of urban air mobility. A couple of months after our first meeting with UDOT and local stakeholders, NASA published the reports that identified key technologies and barriers. In one figure [67], Crown Consulting identified 34 technologies on the critical path of development and divided them into 15 categories: autonomy, sensing, cybersecurity, propulsion, energy storage, emissions, structures, safety, pilot training, certification, communications, controls, operations, traffic management, and infrastructure. This categorization is not to say that these technologies do not depend on each other, and there are complex relationships that must be managed between them during both development and production. Additionally, the airspace is heavily regulated, particularly in the United States where regulations have been developed over the past 100 years; this increases the barrier of entry for innovators due to the capital requirements and consequences of liability. The National Aerospace and Aeronautics Administration (NASA) and the Federal Aviation Administration (FAA) have stepped in to help facilitate the coordination between industry and government, a mission statement from NASA's AAM website (https://www.nasa.gov/aam/overview) provides a concise description of how they see their role:

NASA's vision for Advanced Air Mobility (AAM) Mission is to help emerging aviation markets to safely develop an air transportation system that moves people and cargo between places previously not served or underserved by aviation—local, regional, intraregional, urban—using revolutionary new aircraft that are only just now becoming possible. AAM includes NASA's work on urban air mobility and will provide substantial benefit to U.S. industry and the public.

The Aeronautics Research Mission Directorate (ARMD) initiated the AAM Mission Integration Office during the 2020 fiscal year with the objective to promote flexibility and agility while fostering AAM mission success and to promote teamwork across ARMD projects contributing to the AAM Mission. The AAM Mission will address a broad set of barriers necessary to enable AAM that will be accomplished with the contributions made by projects across the mission directorate.

Aside from governmental players, large corporations with institutional reputations have also stepped in to offer commercial solutions that revolve around an *ecosystem* product concept, a location for providers of services for AAM to market their products.

1.2   Visuals and Concepts

After the first meeting with UDOT, our research group produced a conceptual simulation of airspace corridors over Salt Lake City. One of the issues that we had discussed was the problem of where low-altitude aircraft would fly and how the public concerns about privacy and noise might be addressed. At the state level, the roads are public property, and so the idea was floated that aircraft should simply fly over the roads. Presented as a short video with a camera that rotated over the city, we demonstrated how 3D semi-transparent rectangular corridors could be constructed from geographic information system (GIS) data that was available from state agencies. Although the simulation lacked precise placement of corridors and their merge points, the visualization had the effect of catalyzing more conversation in the state and generating interest from multiple industry stakeholders, including GE (AirXOS), AirMap, and Bell Aircraft.

Airspace visualizations and simulations are a powerful tool to guide conversation and facilitate coordination. Visit any website of the major players, Joby Aviation, Uber, Google, AirBus, etc., and you are bound to find an animated visualization depicting the future of air travel. However, simulations are normally constructed to answer more specific engineering problems, rather than as a marketing tool to generate interest. Our initial visualization was constructed using an open-source library called NASA Worldwind. This is a 3D geospatial visualization library with bindings for Java and web technologies such as Javascript. The video that we shared with UDOT and others was a Java program that read GeoJSON data describing the road network around the University of Utah and then constructed three-dimensional corridors at a fixed height above the ground. A small applet with 3D controls then allowed the user to pan and rotate around the area of interest. A movie was then created by programming incremental rotations about a fixed center and storing frames from the applet at a rate of 15 frames per second (fps). The width and height of these corridors were chosen arbitrarily to make the visualization appealing. Additionally, several spherical objects, representing aircraft, were programmed to "fly" through the corridors.

That same year, NASA announced that it would end support for Worldwind, so our team looked elsewhere for a visualization tool. Most of our rapid prototyping efforts utilize MATLAB, with visualization presented on a generic 3D canvas (using the plot3 function). However, there is also a sense, garnered through many conversations with stakeholders, that to make AAM research palatable to a large audience, it would require more specialized geospatial visualization tools. To this end, a business opportunity arose: a platform for AAM related products could support and accelerate advanced air mobility by making it easier to pitch, develop, test, and deploy research and software technology.

1.3   Technology Opportunity

In the push to adopt Advanced Air Mobility, stakeholders include governmental bodies charged with overseeing airspace utilization, as well as Providers of Services for Urban Air Mobility (PSU), and UAS Service Suppliers (USS). UAS operators such as Amazon, UPS, hospitals, etc. are anxiously awaiting operational Unmanned Aircraft System (UAS) Traffic Management (UTM), which will enable package and drug delivery, as well as unmanned air taxi services. The Global UTM Association defines UAV Traffic Management as a system of stakeholders and technical systems collaborating in certain interactions, and according to certain regulations, to maintain safe separation of unmanned aircraft, between themselves and from Air Traffic Management, at very low level, and to provide an efficient and orderly flow of traffic [43]. Companies such as AirMap, Bell Helicopter, GE, and others have expressed great interest in exploiting such a system. NASA has done market surveys that indicate that by 2030 there may be 750M air taxi flights and 500M package deliveries per year in 15 major cities. In addition, this work may allow efficient integration and synergy between ground and air vehicles. Finally, the existence of such a system will also enable the acquisition of a whole new source of big data (flight data, sensor data, communications data, weather data, etc.) that may form the basis for a wide variety of new services.

Current research and product development aim to catalyze the adoption cycle that underlies the nascent industry of urban air mobility (UAM). In its 2020 forecast publication [37], the FAA acknowledges that "it is extremely difficult to put a floor on the growth of the commercial UAS sector due to its composition and the varying business opportunities and growth paths." However in the same study, they say, "if, for example, professional grade small UAS (sUAS) meet feasibility criteria of operations, safety, regulations, and satisfy economics and business principles and enter into the logistics chain via small package delivery, the growth in this sector will likely be phenomenal;" phenomenal, relative to the forecast of about one million

non-model aircraft operating for commercial reasons in 2024, each registering
multiple flights per day [37]. This fleet does not include the vehicles expected to
deliver about one million express packages in that same year, according to a study
conducted by NASA [67, 78]. The FAA also estimates between 12,000 and 23,000
passenger-carrying autonomous aircraft operating within urban environments by the
year 2030. As the FAA suggested in their assessment however, these estimates rely
on the assumption that UAM technology will be adopted and that efficient concepts
of operations (CONOPS) can be developed.

Consulting reports and conversations with industry stakeholders indicate that
most believe *regulation* to be the highest inhibiting factor to growth of the UAM
industry. However, NASA's own funded study regarding the barriers to adoption
indicates a much more complex landscape, including technical factors as well as
market conditions. Therefore, the more realistic view sees regulation as an outcome
of progress in the technological development of this industry. The more realistic
characterization is where conflicts exist between every pair of stakeholders, and it is
the complexity of these relationships that inhibits growth.

One of the authors, D. Sacharny, has developed the GeoRq platform that
addresses these complexities by providing a collaborative integrated development
environment with specialized system development tools and by structuring the
problem in terms of system-level policies and agent behaviors (see Fig. 1.2)
using the lane-based approach described throughout the book. Three organizational
components form the platform: tools to create and store requirements (specifically
geospatial–temporal requirements), tools to create impact and benchmark metrics,
and tools to create real or simulated deployments. Both the lane-based approach
and the platform are critical components because one provides the conceptual

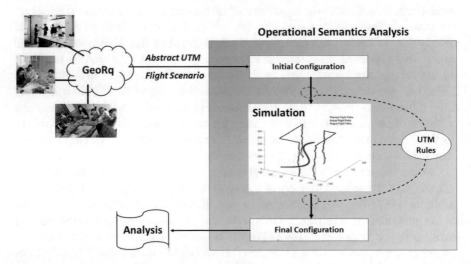

Fig. 1.2 The core component of the GeoRq platform

and computational framework for analysis, and the other provides a vehicle for collaborative engineering and commercialization.

Example Business Model

The main revenue streams for such a product include subscription to cloud services (deployed and secured platform workspaces) and access to APIs and microservices such as the Lane-Based UAS Management System, licensing, and data-access fees. For example, the *GeoRq Workspace* is a cloud deployment consisting of multiple connected instances of virtual machines (VM), databases, and configurations. A GeoRq Workspace may feature an instance of a flight scheduling system, an instance of Eclipse Theia with GeoRq extensions, GeoServer to provide web map services, two instances of GeoRq's PSU, an OIDC security server, and 2TB of Google-backed storage. This setup supports designing, testing, and deploying large-scale logistics operations: one PSU communicates with the region's UTM, while the other forms a digital twin to simulate deployments, and the Eclipse Theia instance with GeoRq extensions serves both the end-user as an Air Traffic Operations Center (ATOC) and the developers as an integrated development environment. Workspace configurations can be updated dynamically with fine-grained resource pricing, and each workspace supports multiple users (contingent upon resource requirements).

In a nascent industry such as UAM, companies must replicate a similar structure of computational instances to conform to UAM system policies. However, the intense competition between current players to develop, and become the standard bearer of UTM software, has forced much of the common architecture into proprietary silos. The result is that non-recurring engineering (NRE) in this space, such as required by new-product development, is expensive and compounds with each new engineer that must climb the same hill.

Open-source development, as with GeoRq, overcomes this problem by packaging up the common architecture, making it configurable, extensible, and deployable, and by providing an integrated open-source systems development tool. Product developers can then repackage proprietary APIs, datasets, microservices, user interfaces, etc., and deploy the white-labeled GeoRq Workspaces as a new product for their clients. Reducing NRE by building products using open-source and collaborative software enlarges the pool of qualified designers, engineers, and users, and it can have dramatic effects on the growth of industries.

In the case of a minimal GeoRq Workspace, not including strategic deconfliction or PSU deployments, a standard software estimation tool applied to the current code base estimates approximately 17 months and 8 engineers to complete this common architecture. The cost estimate of $1.6M$ assumes an average wage of $56,286; however, a higher average wage is likely due to the narrow expertise required.

After many discussions with potential subscribers, industry stakeholders, and government, our observation is that the drive to create products for the UAM industry exists across many disciplines. Table 1.1 shows a sample of the companies interviewed during our research of this problem. For example, a company might acquire a patent for advanced trajectory generation. After integrating the capability into a web-based API, they would spend considerable NRE developing visualiza-

Table 1.1 Stakeholders interviewed

Firm name
Crown Castle
Crown Consulting
Skytelligence
SmartSky Networks
AiRXOS (GE)
UPS
Aerial Transportation Solutions (ATS)
AirMap
ANRA Technologies
University of North Texas
Camel Works Design (Dubai Road Transit Authority)
Anne Arundel Hospital System
Alakaí Technologies
Westinghouse Electric Company
Fortem Technologies
CogniTech Corporation
University of Utah Health

tions using, for example, NASA's WorldWind libraries for marketing purposes. Given the chance to use a tool such as the GeoRq workspace and the visualization capabilities available there, the API strategy might change considerably. The realization would be that packaging a company's technical capability within a platform such as GeoRq provides a powerful channel to market their product as part of a deployable system. Another example would be a developer engaged in the NASA AAM national campaign in order to commercialize communications research. This would require the development of a PSU for a valid simulation and the necessary infrastructure to deploy a production instance of their technology—this is a costly endeavor considering the NRE required. Access to a GeoRq-like system could accelerate their research. Integration of UAM infrastructure would allow product developers across industry to demonstrate the feasibility and potential for commercial investment. It would not be necessary to spend a considerable amount of NRE developing a web-based system for exploring and visualizing their data, including updates for changes in the AAM framework as this industry develops; systems such as GeoRq are a cost-effective alternative.

A viable business model emerged through these discussions: offer product developers a configurable, cloud-deployable package containing the prerequisites for any UAM product. A basic set of features would be included, with additional cloud capabilities and deployments (such as large-scale publish/subscribe frameworks) available through fine-grained resource pricing. An open-source tier is provided to generate community engagement and sustainable commitment to the platform, allowing developers to customize the platform as the UAM industry evolves. An *Individual* tier addresses the needs of smaller firms, individual entrepreneurs, and researchers. The *Enterprise* tier is for firms that plan to develop multiple products or to deliver the white-labeled platform as a product to downstream clients.

To estimate potential revenue given this pricing model, a sample list of potential subscribers was collected from pre-certified consulting firms for several state departments of transportation (U.S. based). The list was narrowed to consulting firms with the following capabilities, having a high likelihood of serving UAM requirements: surveying and mapping, geotechnical services, traffic operations design, traffic engineering and operations studies, and environmental studies. This compiled list included 1383 firms with an estimated median of 32 technical staff per firm. We expect that technical staff will be drivers, as well as end-users, for adopting a platform such as GeoRq. As an example, the total number of technical staff present in one dataset (the most descriptive dataset) was 158,286 people. For this sample of the total addressable market, if 0.3% of the technical staff see potential in serving UAM requirements with their capabilities and each adopts a single enterprise tier package, then the total annual revenue exceeds $28M. This figure considers the first workspace adopted by these developers, and it becomes compounded as more products are developed, white-labeled, and adopted by downstream clients. Furthermore, this sample market represents a fraction of the developers that will enter this industry in the next few years. The total addressable market for a GeoRq-like tool is likely orders of magnitude above this sample, especially if complementary markets (GIS, programming IDEs, cloud computing) are considered.

The margins on selling this type of NRE are large, the marginal cost to run the enterprise tier in the cloud runs annually about $362. For a firm, or even an individual, deciding whether to venture into product development in this nascent UAM industry, the value proposition is dramatic: a GeoRq-like product reduces the necessary investment by at least $2M$ and accelerates development by at least 1.5 years.

Commercialization Approach
The GeoRq platform is an example vehicle for commercializing research. Research efforts produce software to perform simulations, record and validate benchmarks, and test assumptions. Source code can be delivered directly as part of a workspace configuration or wrapped in a microservice. Front-end code is engineered by programmers using GeoRq extensions and then included with individual or enterprise tiers. The commercial feasibility of each product is measured by the value (the marginal price of selecting this feature with a GeoRq workspace) over the cost of the computational resources required to run that feature in the cloud (e.g., required datasets, storage requirements, etc.) and the NRE required to produce it.

Developers of systems such as GeoRq can apply for a variety of assistance from state and local entities to assist with portions of business development and commercialization. It is usually possible to work with the state agencies to identify, bid, and win procurement opportunities with federal, state, and local government entities. Furthermore, it is possible to seek assistance from the appropriate Small Business Development Center (SBDC) to receive business counseling and assistance in business plan development.

Chapter 2
Introduction to UAS Traffic Management

2.1 Introduction

In early 2018, the director of the division of aeronautics at the Utah Department of
Transportation (UDOT) invited our research team to a working group discussion
about enabling Urban Air Mobility (UAM) in the Salt Lake Valley in Utah. At
the time, we had been working on a program sponsored by the Air Force Office
of Scientific Research (AFOSR) to develop a Dynamic Data-Driven Application
System (DDDAS) for Geospatial Intelligence [94, 96].[1] Our research into data
fusion techniques were relevant to UAM since they involved unmanned aircraft
systems (UASs), weather, and decision making. Representatives from industry,
such as Fortem Technologies Inc., raised practical concerns about the positioning
of radar systems for tracking low-altitude aircraft, while urban planners discussed
constraints for zoning, and local government stakeholders addressed public funding
and perception. Over the course of that year, we met with vehicle designers like
Bell Textron Inc., designers of the concept Bell Nexus, and system designers like
General Electric AiRXOS and AirMap. At the same time, NASA and other industry
partners were gearing up for Technical Capability Level-4 (TCL4) flight tests for
enabling small UAS (sUAS) (55 lbs. or less) operations in low-altitude airspace
(typically uncontrolled or Class G airspace under 400 feet above ground level
(AGL)), conducted at a Nevada, USA test site [53, 63]. While the operational
requirements for sUAS (usually small business or hobby use-cases) differ from
medium to large UAS at higher altitudes, the methods proposed by NASA and
the FAA for distributed coördination rest on the same basic architecture, shown
in Fig. 2.1. This architecture describes the roles and responsibilities of agents in the
UTM system, as well as the delegation of authority. However, questions about the

---

[1] This chapter is based in part on enhanced versions of [89].

© The Author(s), under exclusive license to Springer Nature Switzerland AG 2022   11
D. Sacharny, T. C. Henderson, *Lane-Based Unmanned Aircraft Systems Traffic
Management*, Unmanned System Technologies,
https://doi.org/10.1007/978-3-030-98574-5_2

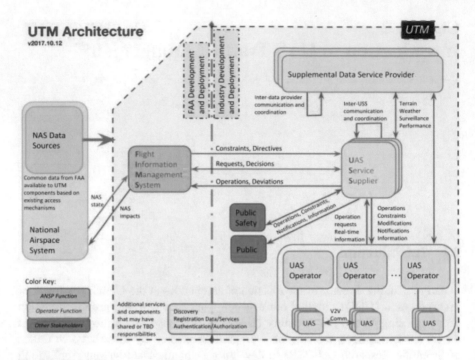

Fig. 2.1 NASA/FAA proposed UTM architecture (from [85])

inherent safety, predictability ,and scalability of the system require a deeper look
into the algorithms and behaviors that define the individual agents.

This book presents a structured airspace for UTM systems to help answer
practical questions about expected system behavior. *The lane-based approach is an
effective organizational strategy because it allows efficient strategic deconfliction as
well as the determination of the impact on contingency analysis and handling due
to the interaction between agent behaviors and UTM policies.* Complex systems
deployed in the real world are bound to experience contingencies: i.e., possible
future events, usually causing problems or making further plans and arrangements
necessary. Planning for contingencies, including conflict management, is a core
computational issue underlying large-scale autonomous systems because computing
optimal plans is an intractable problem (for both software systems and human
operators). Therefore, the interplay between UTM policies (including airspace
structure, communications, etc.) and UAS behaviors, which encode individual
preferences and autonomy, can have a dramatic effect on contingency handling.

The lane-based approach to conflict management structures the airspace with
one-way volumes and special constraints on intersections and represents both a
practical organizational strategy and a tool for analyzing individual and system
behaviors. It is neither completely decentralized nor centralized, allowing system
designers to make explicit compromises between preferences, and design for contin-

gencies. Through agent based modeling and simulation (ABMS), new applications of spatial measures and analytical tools, and a comprehensive review of related research, the effectiveness of this approach will be shown.

2.2 NASA/FAA UTM Background

NASA's proposed architecture for UAS traffic management (UTM), shown in Fig. 2.1, draws inspiration from the current national airspace system (NAS). It is a distributed computing system, where operators are individually responsible for planning their flights, ensuring they do not conflict with any other planned flights, and obtaining the requisite permissions to fly from authorities. The overarching governing authority, represented by the Flight Information Management System (FIMS), delegates authorization to UAS service suppliers (USS),[2] which are certified automated systems that ensure flights are strategically deconflicted[3] before authorized. USS may then authorize flights for multiple operators or vehicles and could represent an organization such as Amazon or Google, which may serve thousands of aircraft, or a smaller entity with one or a few aircraft. The main constraint that this proposed architecture applies is in terms of interfaces between the high-level operating agents (namely the PSUs, operators, and regulators). The sequences of interactions between them are still mostly undefined and flexible. For example, there is no rule that inhibits PSU from designing a structured airspace for its operators and then scheduling them using a proprietary protocol. There is also nothing stopping state regulators from imposing a structured airspace within which all PSU must operate. Consequently, these engineering questions remain open-ended and experimented by NASA, and industry is underway.

In particular, there still remains the question of exactly how safe separation is achieved across all operations at all times. The FAA and NASA have delegated the responsibility of ensuring strategic deconfliction to the USS, but the specific methods for trajectory planning within this system are left unspecified. During several developmental test efforts held by NASA and the FAA, the concept of a Discovery and Synchronization Service (DSS) was invented to address the problem

---

[2] In this book, USS and Providers of Services for UAM (PSU) will be used interchangeably. At the time of this writing, PSU was a relatively new acronym for a functionally similar role as the USS, with specific interface requirements for mid-to-large autonomous aircraft.

[3] Strategic deconfliction, or strategic conflict management, refers to the first of three layers of conflict management defined by the International Civil Aviation Organization (ICAO), "achieved through the airspace organization and management, demand and capacity balancing, and traffic synchronization" [49]. This is generally understood to mean that before an operation takes flight, its planned trajectory does not violate minimum separation constraints with any other planned flight. The next layers are applied in order of the shrinking conflict horizon, and they are *tactical* in nature and termed "separation provision" and "collision avoidance."

of locating other operations within an area. This role is critical to ascertain the state of the airspace and acts as a centralized database. The internal representation of the DSS is not a specific requirement; however, the interfaces are being developed as an industry standard [8]. A prototype DSS has been constructed by Google's Project Wing, called InterUSS [45], that has been used by NASA's developmental efforts. The internal representation of InterUSS utilizes S2 cells at a fixed zoom level (see Fig. 2.2). The unit sphere is decomposed into a hierarchy of cells by a framework called the S2 library, where four geodesics bound each cell of a quadrilateral. The cell hierarchy is created by projecting the faces of a cube onto the unit sphere and then recursively dividing cells into 4 sub-cells. When an operation is planned, the USS queries these cells to find other USSs operating in the area, as well as a unique token to mark the current state of operations in the environment. The other USSs are then contacted to find the operational volumes they have reserved in order to deconflict the planned operation. Once deconflicted, the USS creates a unique reference to the new operation and registers it with the DSS, along with the previously obtained state token. If the airspace state has mutated during trajectory deconfliction (a process that is not standardized), then the DSS rejects the new operation reference, and the USS is required to restart the process. To date, there

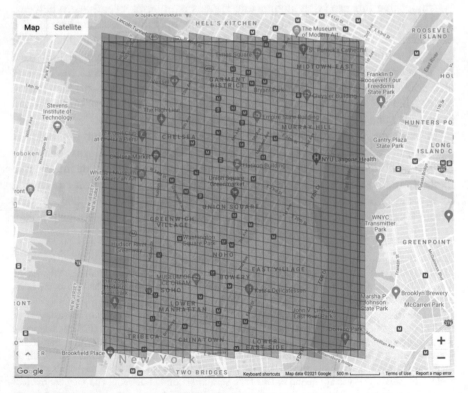

Fig. 2.2 S2 cells covering a portion of New York

is no published work on specific guarantees offered by this approach—one can conceive of a situation in which the environment state mutates so often and so quickly that a particular agent may never be able to fly.

2.3 UTM Scheduling Problem

The online-over-time aspect of the UTM problem means that any algorithm, currently known to computer science, can only be an approximation of an algorithm that produces globally optimal solutions (for the most desired cost formulations, such as maximum delay). In other words, trade-offs between different types of UTM systems are essentially trade-offs between algorithmic heuristics. A number of problem formulations, developed for different application areas, provide a foundation for understanding the heuristic trade-offs, as well as different perspectives represented as quantifiable aspects of the system. The following subsections describe some of the comparable problem formulations.

2.3.1 The Air Traffic Flow Management Problem

A natural problem model comes from research into the *Air Traffic Flow Management Problem* (TFMP) [14]. In this model, the airspace is partitioned into sectors that are controlled by regional regulators who provide separation services. The sectors are characterized by capacities that represent the maximum number of aircraft that may be in a sector at any time, and depend on factors such as weather. TFMP implements two control strategies to ensure that sector capacity constraints are not violated: ground-holding and speed adjustment. Ground-holding shifts the entire flight in time by delaying the departure of an aircraft. Speed adjustment is applied to each sector in flight and represents an "air delay." Optimal ground-hold time and speeds for every planned flight are calculated, but each operation does not deviate spatially (this is called the *Air Traffic Flow Management Rerouting Problem* (TFMRP) [14]).

Rios and Lohn [83] compare techniques for finding a solution to the Bertsimas and Stock-Patterson (BSP) model: binary integer programming, genetic algorithms, and simulated annealing. They also compare a greedy scheduler that schedules flights on a "first-come, first-serve basis by finding the first available departure time for each flight in turn that will not violate sector capacities when combined with previously scheduled flights." The greedy scheduler is so named because it is locally optimal for the flight in question, but it does not guarantee globally optimal solutions. Solutions to the BSP model provide time intervals during which a flight must enter each segment, and the solution is guaranteed to minimize the total delay. Although the size of the problem formulation is bounded by a linear relationship between the number of intervals, the number of flights, and the number of sectors, the integer linear programming formulation suggests that there is no known time-polynomial algorithm to solve it [73].

Despite its non-deterministic features, this representation is appealing because it supports the goals of strategic conflict management, namely "airspace organization" via sectors, and "demand and capacity balancing, and traffic synchronization" via ground-holding and speed adjustment. What it lacks is an explicit representation for the intersection of routes in 4-D space. The primary issue is that the sectors are large, and there is no way to tell if routes intersect. One way to adapt this representation is to shrink the size of each sector such that capacity is fixed to one aircraft per sector. Bertsimas and Patterson explored this assumption and determined that the computational complexity of the TFMP is NP-hard [14]. Also, reducing the size of the sectors dramatically increases the space complexity.

2.3.2   The Job-Shop Scheduling Problem

By shrinking sector capacities to one, the TFMP can be reformulated as a scheduling problem (see [73, 75] for a definition of the general scheduling problem, and [7] and [54] for an overview of the job-shop scheduling problem). Bertsimas and Patterson [14] reformulate the problem as follows: for each job, create an aircraft, and for each processor, associate a sector (sectors include airports). Each job is composed of tasks that represent a flight segment (time spent in a sector). A solution to this formation is a total ordering of sectors for every job, and a list of flight times for each task such that the total delay is minimized, and all flights are performed by a deadline. This formulation guarantees that no aircraft will occupy the same sector at the same time and therefore satisfies the non-intersection requirement.

There are, however, several practical issues with this formulation. To begin with, the job-shop scheduling problem is NP-hard—this makes it a poor choice for USS that may need to contest with tens of thousands of "jobs." Furthermore, it is not clear what the sector size should be, given the variation of UAS sizes expected to utilize the airspace. Too large a sector could result in an unreasonable amount of tactical separation maneuvers, while too small a sector could become computationally intractable. To account for uncertain speeds, the scheduling model can incorporate probabilistic durations. This formulation still suffers from the time complexity as before (and likely worse if the durations are not assumed to be independent random variables) [26].

2.3.3   The Multi-Robot Motion Planning Problem

Strategic deconfliction may be cast as a multi-robot motion planning problem. The key concept for any motion planning problem is the *configuration space*, which combines the kinematic constraints of the robot and the environment. Multiple robots may be combined into a conceptual "composite robot" [119], and the motions are planned in a joint configuration space. Centralized, or coupled, algorithms provide a path for every robot, while decentralized, or decoupled, algorithms usually provide solutions for a subset of the robots. In [119], the authors decompose a multiple robot planning problem into partitions of robots that are planned together. While this approach does reduce the complexity of the joint configuration space, it does not guarantee a reduced complexity of the problem because each partition can still be very complex. Other approaches, such as incremental coördination [102], combine the centralized and decentralized algorithms into a single iteration.

Multi-robot motion planning is also a more natural representation for autonomous vehicle coördination because solution methods, such as *rapidly exploring random trees*, can incorporate dynamic constraints and uncertainty directly. The desire for optimality, however, results in a worse-case time complexity comparable to the job-shop scheduling problem. The two-phase decoupled approach [102] involves first computing a path for each robot individually while ignoring other robots, and then operations are applied to the resulting path set to avoid collisions. The advantage of this approach is that the "search space explored by the decoupled planner has lower dimensionality than the joint configuration space explored by the centralized planner" [102]. The drawback is that it is an incomplete algorithm, meaning it is not guaranteed to find a solution even if one existed by considering the system as a whole. This approach resembles the greedy, first-come, first-serve algorithm described by Rios in the sense that previously planned paths are considered as static obstacles and each new flight is delayed until the capacity constraints are met [84].

2.3.4   The Traffic Assignment Problem

The traffic assignment problem (TAP) is a sub-problem in the transportation planning process that models the route-choice behavior of travelers given a set of possible routes [74]. This problem is mentioned here because prior research such as [22] measures the performance of the airway system by simulating origin-destination data from population centers. When determining the capacity of a particular network configuration, the traffic assignment problem should be considered separately because its benefit is mainly to predict the demand on the system. Solutions to TAP result in aggregate measures, "a macroscopic description or prediction of the traffic volume" [74]. The relationship between volume of travelers and their average travel time is modeled by *link performance* functions [74]. Queuing models also play a role in the development of link performance functions.

2.3.5   The Optimization Problem

The FAA expects tens of thousands of UAS to utilize the airspace in close proximity; therefore, the problem model composition is important to ensure that minimum separation requirements are met. There are two ways in general to represent the safety requirements, as a constraint and as an objective function. The objective is to maximize the separation (or headway) between UASs. Assuming the solution is optimal, the question of whether it meets the safety requirement is determined by a threshold, e.g., "the minimum separation is at least 10 meters," or "the minimum separation is at least 10 meters with 99.9% probability." Given the complexity of the UAS strategic deconfliction problem, we propose to constrain consideration to a linear ordering model, that is, well-separated flow along lanes. Thus, the general 4-dimensional space–time trajectory problem becomes one of the deconflicting flights through a connected set of one-way lanes. This approach is developed in the rest of the book.

Chapter 3
Lane Networks

3.1 Introduction

There are many reasons to fly UAS in an urban environment. Several expected high usage applications are package delivery (e.g., food, medical supplies, general goods), inspection (e.g., buildings, bridges, power infrastructure, etc.), and air taxi service. Major companies such as Amazon, UPS, the Postal Service, etc., may deploy hundreds or thousands of UAS regionally per day. Every one of these UAS will follow some trajectory according to a specified time schedule; this is a 1-dimensional curve in a 4-dimensional space. If every UAS creates an individual and arbitrary 4-D curve, then every pair of trajectories must be checked to ensure safety (i.e., minimal separation at all times) and to meet the strategic deconfliction requirements proposed by NASA and the FAA. Moreover, given thousands of densely located UAS in the air at one time, ensuring safe operation may be too complex to allow for human pilots, and system-wide monitoring may be too complex for human air traffic controllers. This means that to enable large-scale UAS flight coordination flights must be autonomous and scheduled so as to avoid conflicts.

3.2 Lane-Based Urban Airways

An alternative to a set of arbitrary trajectories is to create a pre-defined set of lanes through the air and to require that all UAS flights follow these lanes. Each flight must consist of a set of lanes; this starts with a launch lane that takes the UAS from the ground to the air, followed by a sequence of lanes through the air, and terminating with a landing lane that takes the flight from the air to the ground. To ensure safety,

D. Sacharny, T. C. Henderson, *Lane-Based Unmanned Aircraft Systems Traffic Management*, Unmanned System Technologies,
https://doi.org/10.1007/978-3-030-98574-5_3

a time slot through each lane (i.e., a lane entry time and a lane exit time) must be reserved for each flight so that at no time are any two flights too close. This is called strategic deconfliction, and an efficient method for lane-based networks is provided in the next chapter. To make such a method possible, some constraints are placed on the network:

- Each lane is one-way (i.e., the network is a directed graph).
- The direction of an edge is related to compass heading (except those in a roundabout that are always counterclockwise).
- Each lane is at least as long as the headway distance.
- No two disjoint lanes are within headway distance.
- Either the in-degree or out-degree of every vertex is less than two.

Multi-Altitude Airways

An easy way to obtain the basic layout of the airways over a given urban area is to begin by defining an undirected graph at ground level. For example, a simple grid may be affixed to ground locations, or the existing ground road network may be used, wherein every road intersection or termination point is a vertex, and road segments between vertexes are edges. To achieve travel in both directions between air vertexes corresponding to a ground vertex, the air lanes must be placed either side by side at the same altitude above the ground level, or one above the other. The convention used here is that lanes with travel in opposite directions will be vertically separated; moreover, travel in directions $[0, \pi)$ will be at one altitude and in directions $[\pi, 2\pi)$ in the other.

To implement this, there are two levels of airways. In addition, a roundabout is created at each level above a ground vertex. Lanes to enter the airways from the ground are called launch lanes and connect a ground location to the lower altitude airway level. A landing lane connects the lower altitude airway level to a ground location. Both of these ground locations are near the ground network vertex. To achieve these connections, a vertex is placed in the roundabout to connect to a corresponding launch or lane location, respectively.

Figure 3.1 shows a 2×2 ground grid network (an undirected graph with four nodes and four edges) and the corresponding air network (a directed graph with 40 nodes and 56 edges). Note that there are lanes connecting the two roundabouts (separated by altitude) at a vertex location—one up and one down. This allows a flight to enter an air vertex at either altitude and exit along any lane leaving the vertex. Each ground vertex has an associated distinct ground location for its launch and landing lanes, if they exist. As can be seen in the figure, these launch and lane ground vertexes connect directly to corresponding vertexes in the low altitude roundabout above the ground vertex.

The lane-based approach defines a set of one-way lanes where each lane is defined by an entry point, an exit point, and a one-dimensional curve between the two (here we use straight line segments). UASs travel in three dimensions, and thus lanes are understood to be virtual 3D corridors (e.g., cylindrical-like tubes). The shape of corridors may change dynamically and should be constructed to account for the idiosyncrasies of the vehicles that they are meant to support; for

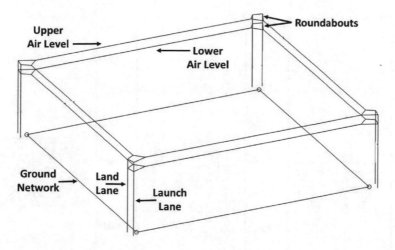

Fig. 3.1   A simple 2 × 2 grid network

example, smaller aircraft in windy environments may require a larger corridor radius than a heavier vehicle with better control dynamics. Further design constraints can be defined in terms of the headway—or safe separation distance—between UAS. The combination of headway and corridor design can support a range of vehicle trajectory constraints, while the directed graph (digraph) imposed on the airspace presents agents with a structured environment for computation (the lanes represent a complete model of the airspace under ideal conditions). Lanes may also have other associated properties (e.g., speed restrictions) specified by the UTM, enabling regulators to communicate requirements effectively to all agents in the system.

Lanes are connected, so that every vertex has either in-degree or out-degree equal to one (except launch lanes that have in degree 0 and land lanes that have out-degree 0). This permits scheduling to be based on lanes as opposed to vertexes since all flights may be deconflicted based on one incoming or outgoing lane, and simplifies the analysis of congestion because various graph-based measures can be utilized to determine most likely high congestion parts of the network. This contrasts with zone-based deconfliction that presumes vehicles can enter and exit in any direction, and the entire zone must be reserved (inefficient for large areas), and cell-based deconfliction that combines zone reservation with general motion planning within each cell (similar to the two-phase decoupled approach in [102]). The choice of the lane spatial layout is key to operational performance. As previously described, several alternatives exist:

1. Airways modeled from ground road networks
2. Regular grid networks
3. Networks with specific properties (e.g., Delaunay networks)

We now give a more detailed account of the lane creation process. Lane creation starts with a ground network defined as a graph, $G = (V, E)$, where V is a set of

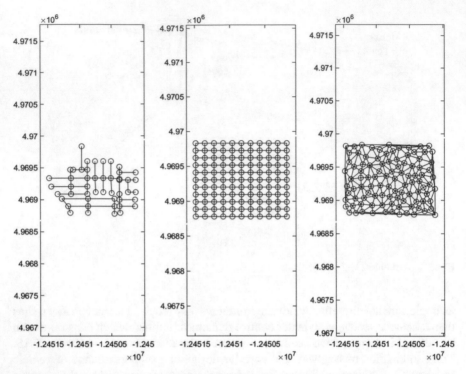

Fig. 3.2 Three types of road layouts over the same locale: actual San Francisco roads (left); grid layout (middle); Delaunay triangulation (right)

ground position vertexes, and E is a set of undirected edges between the vertexes. Figure 3.2 shows example road, grid, and Delaunay networks for a small set of roads from San Francisco, CA. The grid and Delaunay vertexes are within the same geographic area; the Delaunay vertex locations are randomly generated. As will be shown later, the type of network impacts the spatial network measures of the graph. To create the two-level airways between vertexes, the ground network is duplicated as a set of airway lanes at two altitudes: one for travel in direction $[0, \pi)$, and the other in direction $[\pi, 2\pi)$. Since ground vertexes are road intersections, each is represented by two roundabouts in the air centered over the vertex; these ideas are demonstrated in Fig. 3.3.

Consider now, the creation of a roundabout for a vertex p with neighbors $Q = \{q_i | i = 1 \ldots n\}$. A set of vertexes will be placed on a circle of radius r with center p, where r is chosen so that lanes connecting the vertexes on the circle will be at least headway distance, h, apart. A vertex, v_i, is created for every neighbor where the line between p and q_i intersects the circle. To determine the value for r, consider Fig. 3.4. Let d be the distance between two consecutive points on the roundabout. Then:

$$d^2 = r^2 + r^2 - 2r^2 cos(\theta)$$

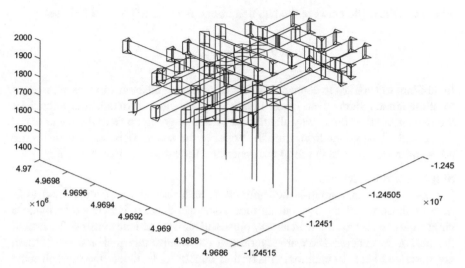

Fig. 3.3 An example two-level grid lane layout of San Francisco roads

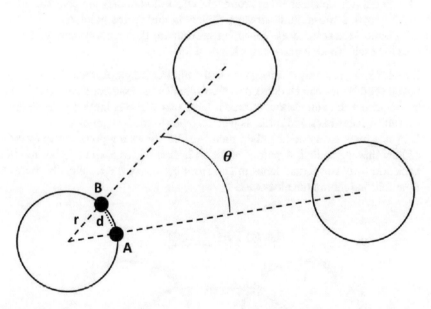

Fig. 3.4 Geometry for determination of minimum required radius. *A* is the point of intersection of the circle and the line between the vertex and one neighbor; *B* is the point of intersection of the circle and the line between the vertex and the other neighbor; r is the circle radius; θ is the angle between the neighbors

where θ is the angle between the two neighbors. In order to obtain $d \geq h$, set

$$r = \frac{h}{\sqrt{4cos(\theta)}}$$

In addition to vertexes to connect to other vertexes, roundabout vertexes are needed to allow launch and/or land lanes. Finally, since there is a roundabout at each of the two altitudes of the airway, there must be a lane going up from the lower to the upper, and a lane going down from the upper to the lower. These extra vertexes are added midway between the neighbor generated vertexes (e.g., like A and B).

Single Altitude Airways

For some purposes, a multi-altitude airway is undesirable; e.g., people in air taxis may be discomfited by frequent altitude changes. Thus, it is better to provide a single altitude airway. Such an airway must still satisfy the lane constraints given at the start of the chapter. There are several ways to achieve this goal, and two of them are described here. In addition, one further constraint is added: the digraph must be strongly connected (except for landing lanes that terminate on the ground); this means that there must be a path from every vertex to every other vertex.

The first option considered uses roundabouts, although they are constructed in a different way to multi-altitude airways. Given an undirected graph consisting of a set of ground vertexes and the edges between them, then a set of new vertexes is created for each. There are two cases to consider:

1. If a vertex, v_1, has only one neighbor in the undirected graph, then Fig. 3.5 shows how a set of lanes can be defined, which satisfies the lane constraints. This does not require a full roundabout but must have two vertexes around each neighbor that allow lanes back and forth, as well as lanes to the interior of v_1.
2. If v_1 has more than one neighbor, then Fig. 3.6 shows a representative example of how this is handled. A pair of vertexes is created for each neighbor to allow back and forth travel, and lanes to and from v_1. All of the local cycles between v_1 and its neighbors run clockwise.

One Neighbor

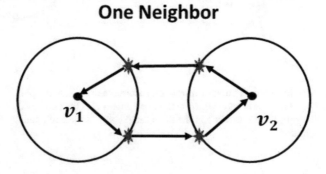

Fig. 3.5 Single altitude lane connections for a vertex with one neighbor

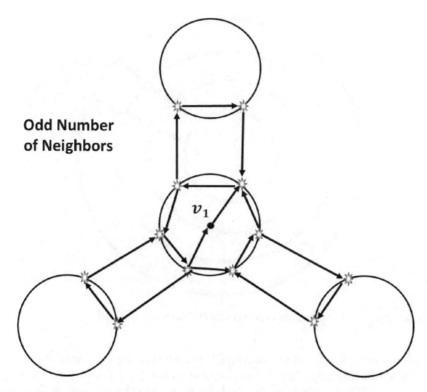

Fig. 3.6 Single altitude lane connections for a vertex with more than one neighbor

This method corresponds to the semantics of roads on the ground in that UASs travel in opposite directions in neighboring lanes in the same horizontal plane. The safety of such a structure requires adequate control of the UAS to avoid crossing into the neighboring lane. Another complication of this approach is the necessity to introduce new vertexes.

An alternative approach is to take the given undirected graph, $G = (V, E)$ as the starting point, and to define a strongly connected directed graph, $G' = (V', E')$, where $V' = V \cup V_{\mathcal{L}}$ and E' are defined below. $V_{\mathcal{L}}$ is a set of ground vertexes for launch and land lanes, which are the only required additional vertexes. One way to achieve this is as follows. Assign $P \leftarrow V'$ be the set of points under consideration. Initialize $E' = \emptyset$. Next, set $k = 1$ and find the set of points, H_k, in the convex hull of P given in counterclockwise order; define a set of edges, $E_k = \{(p_i, p_{i+1}) \mid p_i$ and p_{i+1} are consecutive points in $H_k\}$. Then define $R_k \leftarrow (H_k, E_k)$, as the kth ring for the airway. Now, set $P \leftarrow P - H_k$ and $k \leftarrow k + 1$ and repeat the process so long as $P \neq \emptyset$. This will produce a set of convex polygon rings with a path from every node in the ring to every other node in the ring. Next, a set of edges must be defined to connect the rings. The basic idea behind this is to run a ray from the center of the rings out and create an edge between the pairs of neighboring rings using the

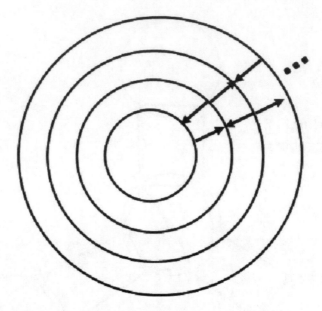

Fig. 3.7 Rings must be connected so as to satisfy the lane constraints

points where the line intersects the rings. There are three separate cases to consider:
the innermost ring may have (1) three or more points (i.e., it is a regular ring), (2)
two points, or (3) a single point. In case 1, the mean of the locations of the innermost
ring points is used to create the line. For case 2, the two points need to be connected
by an edge—call it (p_1, p_2); then p_2, the head of the edge, is connected to the next
outer ring, and some point on the next outer ring is connected to the p_1, the tail
of the edge. If there is a single point, then an edge is created from the next outer
ring to it and an edge from it to the next outer ring. The edges connecting the rings
must maintain the lane constraints, and in particular, every node must have either
in-degree or out-degree less than two; Fig. 3.7 shows how this can be accomplished.
The algorithm is:

Algorithm convex hull single altitude airway
$P \leftarrow V'$
$E \leftarrow \emptyset$
$k \leftarrow 0$
while $P \neq \emptyset$
    $k \leftarrow k + 1$
    $E_k \leftarrow \emptyset$
    $H_k \leftarrow$ convex hull of P
    $E_k \leftarrow E_k \cup \{(h_i, h_{i+1}) \mid h_i, h_{i+1} \in H_k$ are consecutive points$\}$
    $P \leftarrow P - H$
end while
$maxRings \leftarrow k$

if $|H_k| > 2$

    $p \leftarrow mean(H_k)$

    $line_1 \leftarrow$ line through p and $[p_x - p_y]^T$

    $\mathcal{I}_+ \leftarrow$ intersections of $line_1$ with E_k in $\frac{\pi}{4}$ directions

    $\mathcal{I}_- \leftarrow$ intersections of $line_1$ with E_k in $-\frac{\pi}{4}$ directions

elseif $|H_k| == 2$

    $p_1 \leftarrow$ first point in H_k

    $p_2 \leftarrow$ second point in H_k

    $E \leftarrow E \cup \{(p_1, p_2)\}$

    $line_1 \leftarrow$ line through p_1 and p_2

    $\mathcal{I}_+ \leftarrow$ intersections of $line_1$ with E_k in $\frac{\pi}{4}$ directions

    $\mathcal{I}_- \leftarrow$ intersections of $line_1$ with E_k in $-\frac{\pi}{4}$ directions

    $E \leftarrow E \cup \{(p_2, q_1)\}$, where q_1 is an appropriate intersection point with H_{k-1}

    $E \leftarrow E \cup \{(q_2, p_1)\}$, where q_2 is an appropriate intersection point with H_{k-1}

elseif $|H_k| == 1$

    $p =$ point in H_k

    $line_1 \leftarrow$ line through p with direction $\frac{\pi}{4}$

    $\mathcal{I}_+ \leftarrow$ intersections of $line_1$ with E_k in $\frac{\pi}{4}$ directions

    $\mathcal{I}_- \leftarrow$ intersections of $line_1$ with E_k in $-\frac{\pi}{4}$ directions

    $E \leftarrow E \cup \{(p, q_1)\}$ where q_1 is appropriate intersection point in H_{k-1}

    $E \leftarrow E \cup \{(q_2, p)\}$ where q_2 is appropriate intersection point in H_{k-1}

end

$E \leftarrow E \cup \{(q_i, q_j)\}$ where q_i, q_j are intersection points on the ith and jth rings

Using this method, a single altitude airway may be constructed (without round-abouts), which is strongly connected. Figure 3.8 shows an airway developed from a set of vertexes from a grid, while Fig. 3.9 shows an airway developed from a set of random points.

3.3 Spatial Network Measures

Given a specific airway network of lanes, it is important to be able to measure the effectiveness and efficiency of the network. To this end, a variety of measures have been developed (see [13] for a detailed discussion of spatial network measures for road, railway, and other ground networks). Two types of measures are of interest: *static* and *dynamic*. Static measures depend directly on the structure of the graph and the spatial layout of the nodes. For example, a star graph (one node in the center with edges to a set of nodes distributed around it) has a strong point of congestion at the central node since all paths must go through it. On the other hand, dynamic properties (e.g., spatial flow—the number of UASs passing a given point per time unit) depend not only on the network structure, but also on the job mix of the flights

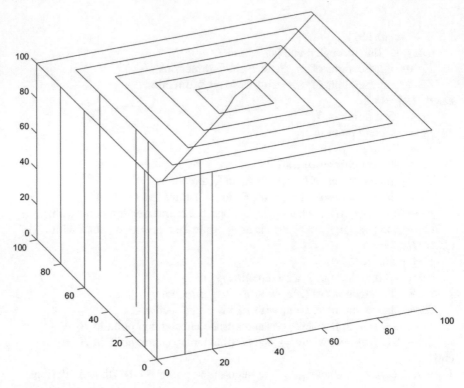

Fig. 3.8 A single altitude airway developed from a set of grid points

and the scheduling algorithm. Generally speaking, static properties are useful when defining the network, and dynamic properties help determine the UTM policies and scheduling algorithm parameters.

3.3.1 Static Spatial Network Measures

Static spatial network measures have been defined to evaluate the quality of a given (ground) transportation network (see [13, 74, 100, 114] for a detailed set of measures), and a set of flow measures (see [100]) as well. A subset of these have been selected to analyze the various road networks used as the basis for airways. A number of spatial network measures have been used to evaluate the quality of ground transportation networks, and they can be applied to airway networks as well. If a ground road network is the starting point for the creation of an air network, then the measures can be applied to the ground network for initial consideration and then applied to the subsequent air network as well. The results on these seem well

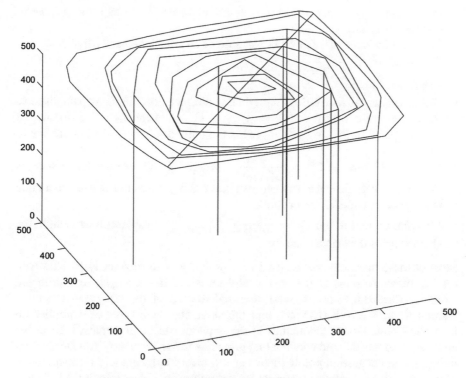

Fig. 3.9 A single altitude airway developed from a set of random points

correlated. Note that, in general, the ground network is a planar, undirected graph, while the air network is a 3D directed graph.

Let $G = (V, E)$ be an undirected graph with $|V| = N$; then the set of static spatial network measures used here includes:

- *Density*: $|E| / |N|$; the ratio of the number of edges to the number of vertexes
- *Total Length*: $\sum_{i \neq j} d_{ij}$, where d_{ij} is the distance from node i to j; i.e., the sum of all edge lengths
- *Minimum Path Length*: (in steps) $L(i, j)$; minimum number of edges from vertex v_i to vertex v_j
- *Minimum Path Distance*: $D(i, j)$ (Euclidean distance); minimum distance traveled along the edges from vertex v_i to vertex v_j
- *Graph Diameter*: $max_{i,j} L(i, j)$; longest path in graph
- *Cyclomatic Number*: $\tau = |E| - |N| + 1$; the number of reduced circuits in the graph
- *Meshedness (also called α-index)*: $\frac{\tau}{2N-5}$; the number of cycles vs. total possible
- *Density (also called γ-index)*: $\frac{|E|}{3|N|-6}$; percent of existing routes to total possible routes
- *Organicness (also called r_n)*: $\frac{N(1)+N(3)}{\sum_{k \neq 2} N(k)}$; $N(k)$ is the number of vertexes with degree k

- *Route Factor*: $Q(i, j), i \neq j$; for each minimal path from i to j, the number of steps over distance
- *Minimum Spanning Tree (MST) Cost*: sum of all edge lengths of MST in G
- *Overall Cost*: $\frac{\text{Total Length}}{\text{MST Cost}}$
- *Efficiency*: $\frac{\sum_{i,j}(\frac{1}{L(i,j)})}{N(N-1)}$
- *Detour Index*: pairwise ratio of straight line distance over minimum path distance
- *Betweenness Centrality*: $bc(v) = \sum_{s \neq v \neq t} \frac{\sigma_{st}(v)}{\sigma_{st}}$, where σ_{st} is the number of shortest paths from s to t and $\sigma_{st}(v)$ is the number of shortest paths from s to t through v
- *Closeness Centrality*: $C_i^C = \frac{N-1}{\sum_{j=1,i \neq j}^{N} d_{ij}}$, where i is the vertex index, and d_{ij} is the shortest path distance between vertexes i and j; measures how close a node is to the other nodes in the network
- *Straightness Centrality*: $C_i^S = \frac{1}{(N-1)} \sum_{j=1,i \neq j}^{N} \frac{d_{ij}^{Eucl}}{d_{ij}}$; captures how straight the shortest paths through a node are

Most of these measures can be used in a similar way to road analysis. However, the last three are some of the most useful measures (for example, see the recent work of Ahmadzai et al. [1] who proposed the use of the Integrated Graph of Natural Road Network (IGNRN) and measured the three types of centrality on it to show how various hierarchies in the network can be determined). Even the application to the 2D road network that gives rise to the 3D airway provides insight into possible congestion points (betweenness centrality), centrally located launch centers (closeness centrality), and nodes through which fewer turns (due to lanes) are necessary (straightness centrality). In addition, these measures provide a visual mechanism to compare different network layouts. For example, Fig. 3.10 allows comparison of grid and Delaunay layouts: brighter nodes are likely more congested. Also, note that the grid network betweenness values are about double those of the Delaunay network.

Table 3.1 gives the values for some of these measures for three Salt Lake City, Utah graphs; the values are indicative of the performance using the graph. In particular, the betweenness centrality (BC) is useful in locating bottlenecks. Now consider the betweenness centrality of the airways over the East Bench area of Salt Lake City, UT. Figure 3.11 shows this measure for the airway constructed from the underlying road network. Figure 3.12 shows the betweenness centrality for the grid network, and Fig. 3.13 displays that of the Delaunay triangulation.

These measures provide clear insight into how the graph affects performance; the Detour Index can help a user select a path, and the last three provide useful information about congestion and flow through the graph. A high measure of betweenness centrality (BC) indicates that a node is prone to congestion since many shortest paths pass through it; high closeness centrality reveals a good site for a launch or land site since the node is close to many nodes; finally, straightness centrality means that shortest paths through this node do not require many turns, which can be important for UAS platforms. For example, Figs. 3.11, 3.12, and 3.13 show the BC measure for three lane networks, and high BC measure corresponds to higher congestion parts of the network.

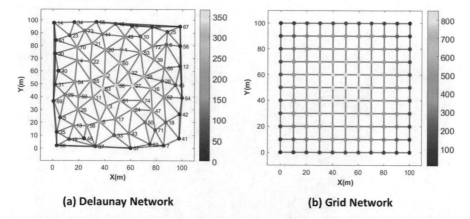

(a) Delaunay Network

(b) Grid Network

Fig. 3.10 (**a**) The betweenness centrality of the grid network; (**b**) Delaunay network (right). Note that the nodes in the center of the grid are likely to be pinch points

Table 3.1 Some measures for the three network types used in the earthquake scenario simulation

Measure	Delaunay	Grid	GIS
Density	2.95	1.91	1.23
Total Len	31.73	32.59	31.01
Graph Diam	30	45	73
Cyclomatic No.	978	483	109
Meshedness	0.98	0.46	0.12
r_n	0.006	0.160	0.786
L_T	4.5915e5	2.525e5	1.006e5
L_MST	9.1041e4	1.3175e5	7.7962e4
Cost	5.04	1.92	1.48
Efficiency	0.1074	0.0840	0.0642
Detour Index	0.94	0.80	0.73
Betweenness Cen.	See graph	See graph	See graph
Closeness Cen.	See graph	See graph	See graph
Straightness Cen.	See graph	See graph	See graph

3.3.2 Dynamic Spatial Network Measures

We have also defined a number of network flow measures (adapted from a standard ground transportation framework—see [74, 100, 114]). These include the *max UAS per lane*:

$$n_{max} = \lfloor \frac{d}{h_d} \rfloor$$

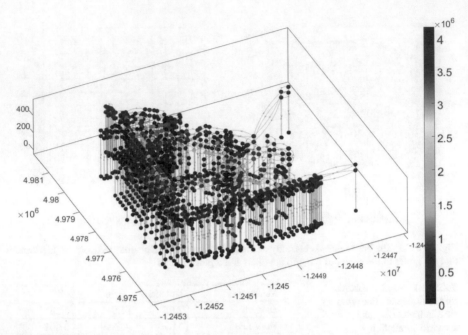

Fig. 3.11 The betweenness centrality of an airway created from the ground road network

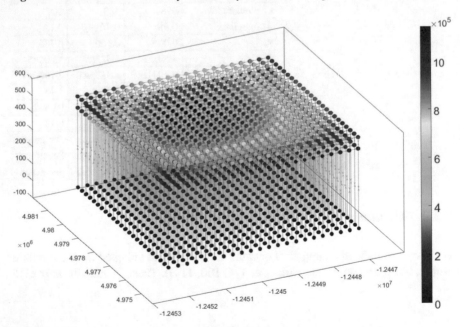

Fig. 3.12 The betweenness centrality of an airway created from a grid

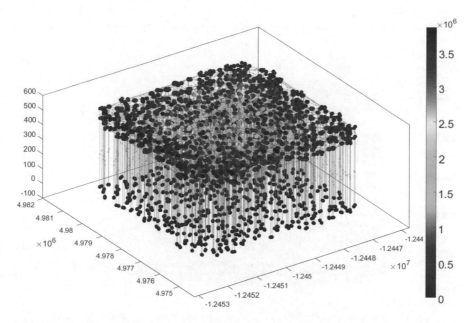

Fig. 3.13 The betweenness centrality of an airway created from a Delaunay triangulation of a set of random points above the same area

where d is the length of the lane, and h_d is the headway distance; the *time occupancy* of a lane:

$$\Theta = \frac{n_k}{t_{max}}$$

where n_k is the total number of UASs through the lane, and t_{max} is the max time considered; *spatial occupancy* per lane:

$$k_s = \frac{\mu}{d}$$

where μ is the average number of UASs in the lane per time unit; and *spatial flow*:

$$q_s = k_s \cdot s$$

where s is the average speed of the UAS.

For the simulations performed here, most of these measures are low given the traffic patterns for supply delivery; however, one interesting result is time occupancy shown in Fig. 3.14. As can be seen, this measure correlates rather directly with the betweenness centrality of the spatial network.

The behavior of requests and the strategy for scheduling can have a significant impact on the average density of lanes. Consider a single lane system of length

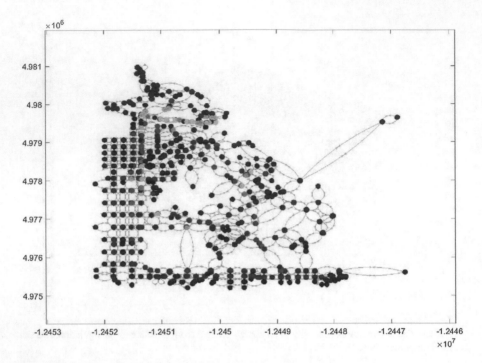

Fig. 3.14 The time occupancy measure for the supply delivery scenario

x, with one entry and one exit. Further assume that vehicles consume a one-unit spatial interval within the lane, and requests arrive over time independently for a uniformly random unit interval. In the first scenario, assume that each vehicle either obtains the requested reservation or drops out, a "failure." This scenario mirrors a 1-dimensional sequential interval packing problem, also known as Renyi's parking problem [80]. Renyi showed that as the length of the lane approaches infinity, the mean filling density approaches 0.7476. This property also holds for the lane scheduling approach given here.

The dynamic measures described above may be used to predict the behavior of lane networks and aid in the design of structured airspaces. However, to compare the lane-based approach with unstructured airspace approaches, such as that proposed by the FAA and NASA, some direct metrics can include:

- *Delay Time*: absolute difference between the desired and actual launch times
- *Deconfliction Time*: wall-clock time required for deconfliction
- *Failures*: the number of flights that could not be scheduled due to conflicts.

Chapter 4
Strategic Deconfliction

4.1 Introduction

The Federal Aviation Administration (FAA) and NASA have provided guidelines for Unmanned Aircraft Systems (UASs) to ensure adequate safety separation of aircraft and, in terms of UAS Traffic Management (UTM), have stated [82]:

> A UTM Operation should be free of 4-D intersection with all other known UTM Operations prior to departure and this should be known as *Strategic Deconfliction* within UTM ... A UTM Operator must have a facility to negotiate deconfliction of operations with other UTM Operators ... There needs to be a capability to allow for intersecting operations.

The latter statement means that UTM Operators must be able to fly safely in the same geographic area. The current FAA-NASA approach to strategic deconfliction is to provide a set of geographic grid elements and then have every new flight pairwise deconflict with UTM Operators with flights in the same grid elements. Note that this imposes a high computational burden in resolving these 4D flight paths and has side effects in terms of limiting access to the airspace (e.g., if a new flight is deconflicted and added to the common grid elements during a proposed flight's analysis period, then the proposed flight must start the deconfliction analysis all over).

We have proposed a lane-based approach to large-scale UAS traffic management [90, 91] that uses one-way lanes, and roundabouts at lane intersections to allow a much more efficient analysis and guarantee of separation safety.[1] We present here the results of an in-depth comparison of FAA-NASA strategic deconfliction (FNSD) and Lane-Based Strategic Deconfliction (LBSD) and demonstrate that FNSD suffers from several types of complexity that are generally absent from the lane-based method.

---

[1] This chapter includes portions of [98, 99].

© The Author(s), under exclusive license to Springer Nature Switzerland AG 2022   35
D. Sacharny, T. C. Henderson, *Lane-Based Unmanned Aircraft Systems Traffic Management*, Unmanned System Technologies,
https://doi.org/10.1007/978-3-030-98574-5_4

Small Unmanned Aircraft Systems (sUASs) are to be integrated into the low-altitude (Class G) airspace, and initial concepts have been provided by the NASA UAS Traffic Management (UTM) project [85]. A set of four Technical Capability Levels (TCL) have been defined, and TCL 4 addresses "an urban environment and includes handling of high density environments, large-scale off-nominal conditions, vehicle-to-vehicle communications, detect-and-avoid technologies, communication requirements, public safety operations, airspace restrictions, and other related goals." Figure 2.1 shows the UTM framework proposed by NASA. The UAS Service Suppliers (USSs) provide key functions for managing the airspace, and in particular, they are charged with ensuring strategic deconfliction (SD) of flights, and other services usually provided by the Air Navigation Service Provider (ANSP) in manned aviation. In addition, USSs are charged with monitoring flight operations. All of these functions are to be achieved in a distributed, coöperative manner. NASA's vision is that USSs perform SD by making sure that any proposed flight has no 4D (space and time) conflict with any scheduled flight in its area of operation. This also permits arbitrary flight paths.

The FAA-NASA SD approach has some glaring problems, including the computational complexity of arbitrary 4D path planning, as well as its susceptibility to monopoly control by organizations with large-scale resources. We have proposed an alternative approach using well-defined lanes [90, 91] in which a (generally) fixed set of lanes (airways) are established, and then flights are scheduled through these lanes. This means that SD becomes a 1D problem that is solved much the same as with manned flight, that is by delay of the takeoff time.

We provide the first detailed set of experimental results that allows analysis and comparison of the two alternative approaches. The results indicate that the lane-based approach is superior in most aspects.

4.2 FAA-NASA Strategic Deconfliction

The FNSD is based on a gridded approach in which the area of flight operations is divided into a number of grid elements, and each flight scheduled by a USS keeps track of the grid elements over which it operates. Then when a new flight is being scheduled, it only needs to deconflict with the flights with which it has common grid elements. The current approach proposed by the FAA/NASA is shown in Fig. 4.1, where USS_1 and USS_2 have a number of scheduled flights (USS_1 flights in green and USS_2 in blue). These flights have already been deconflicted for operation over some time interval. USS_3 wants to schedule a flight (red dashed line), but in order to do so, must deconflict flights pairwise with both USS_1 and USS_{12}, which means all flight paths must be shared. If some other USS manages to deconflict and schedule a flight in this space–time before USS_3, then USS_3 must start the process all over from the beginning [45]. Note that thousands of flights a day are envisioned, thus making the complexity of strategic deconfliction overwhelming. Figure 4.2 shows an example grid layout with a 4×4 set of grid elements (i.e., about 1320 feet on

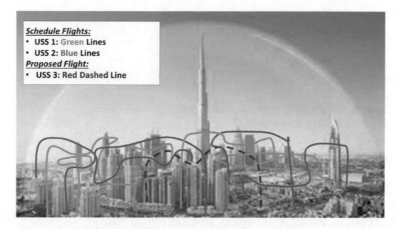

Fig. 4.1 Sketch of the current FAA-NASA proposed strategic deconfliction

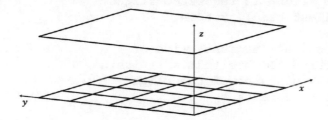

Fig. 4.2 The basic experimental layout with a 4 × 4 grid

a side). For FNSD analysis, we consider a flight path to consist of a polyline with three line segments:

1. *Segment 1*: $[pt_1, pt_2]$, a segment going straight up from a launch site to a chosen altitude
2. *Segment 2*: $[pt_2, pt_3]$, a segment going across the workspace at a fixed altitude
3. *Segment 3*: $[pt_3, pt_4]$, a segment going straight down to the ground

Each flight path is comprised of a user selected launch site, land site, and flight altitude. In addition, each flight has a designated start time and fixed speed for the entire flight. Note that this may be relaxed to allow mixed speeds without much modification of the structure of the deconfliction procedure.

Given a set of flights, the convention is that they are requested and deconflicted in the order of the list. That is, the first flight is scheduled as specified since there are no flights scheduled before it, the second flight must only deconflict with the first, etc. The deconfliction strategy used is based on ground delay of the flight until it has no conflicts with scheduled flights in its grid elements; this allows a fair comparison to the lane-based method that is also based on setting a conflict free launch time; moreover, this is the way standard air traffic control is accomplished. The FNSD algorithm used here is:

```
On input: scheduled_flights,
          flight_request,
          delta_t,
          headway_distance
On output: new_flight

if flight_request shares no grid
   elements with scheduled_flights
then new_flight is requested flight
   (earliest time); return

if there are no flight segments in the
   scheduled_flights within headway
   distance of the flight_request
   segments
then new_flight is requested flight
   (earliest time); return

pinch_pts = all segment pairs of
   scheduled flights that are within
   headway distance of the flight
   request segments

while (any pinch point segments have the
   two flights within headway distance
   during their traversal of the segment)

     shift the start time of the flight
end
```

Although this is just an example of a deconfliction method, the statistics accumulated will be somewhat independent of the particular method. This is due to the fact that the complexity is related to the number of scheduled flights that share grid elements with the flight request and the nature of the segment interactions in the grid element. This method first eliminates from consideration any scheduled flight that shares no grid elements. Next, it eliminates any that share grid elements, but none of whose flight segments are within headway distance of the flight request segments. Finally, to determine whether segment pairs that are within headway distance actually pose a problem, the time of passage of the two flights must be considered. For example, if the entry–exit time through the scheduled flight pinch segment does not overlap the entry–exit time interval of the flight request segment, then there is no conflict. Finally, if these time intervals overlap, then an analysis is performed to see if the flights get within headway distance while crossing their respective segments; if so, the start time for the flight request is delayed a fixed amount, and the impact re-analyzed.

The pinch point segments are determined as follows. Let $P_0 \in \Re^3$ and $P_1 \in \Re^3$ be the endpoints of segment 1, $s \in [0, 1]$, and Q_0 and Q_1 be those of segment 2. Define

$$P(s) = P_0 + s(P_1 - P_0) \tag{4.1}$$

$$Q(t) = Q_0 + t(Q_1 - Q_0) \tag{4.2}$$

$$w(s, t) = P(s) - Q(t) \tag{4.3}$$

Then the distance squared between $P(s)$ and $Q(t)$ is

$$\mid w(s, t) \mid^2 = w(s, t) \cdot w(s, t) \tag{4.4}$$

When the distance is less than the allowed headway, then the pair of segments is recorded.

4.2.1   A Detailed Analysis of Strategic Deconfliction

As described above, a framework is being developed to support large scale (thousands) of UAS flights per day over urban areas, and NASA has proposed a UAS deconfliction strategy that requires service providers (UAS Service Suppliers or USSs) to exchange full flight path information and to mutually find a deconflicted set of flights. This approach has high complexity and sacrifices UAS operator privacy. We propose a lane-based deconfliction strategy that reduces the shared information to be simply lane entry and exit times and UAV speed through the lane. Then given a requested launch time interval, it is possible to determine the set of all allowable (deconflicted) time intervals within the requested interval.

Techniques proposed for flight planning include full mix and layered methods [106, 108] for which safe separation is maintained by tactical collision avoidance methods in otherwise unconstrained flights. While several heuristic methods have been developed for this problem (e.g., [11]), it is still possible that the number of conflicts may overwhelm the algorithms (see [52] for an analysis of cascading effects of conflict resolution). There has been a large amount of research into quantifying the risk of conflict in this type of system (e.g., [16, 22, 52, 108, 113]), indicating that there are numerous risk factors that an operator would need to consider in order to reduce the risk of collision. Lane-based airways were analyzed in [51]; however, the UAS operations were not deconflicted pre-flight and instead were simulated much like car-following models (e.g., [70]). Recently, a report published by NASA detailed the negotiations among stakeholders regarding requirements for USS SD. Furthermore, they discuss the requirement that any scheme for strategic deconfliction must be mandated by the airspace regulator.

The *Strategic Deconfliction Problem* is to produce a set of scheduled flight paths such that no two aircraft ever get closer than a specified safety distance (specified either in space or time).

Strategic deconfliction, or strategic conflict management, refers to the first of three layers of conflict management defined by the International Civil Aviation Organization (ICAO), "achieved through the airspace organization and management, demand and capacity balancing, and traffic synchronization" [49]. The next layers are applied in order of the shrinking conflict horizon and are *tactical* in nature and termed "separation provision" and "collision avoidance." Broadly speaking, strategic conflict management deals with planning collision-free paths, which in the most general case of planning for multiple agents is P-SPACE-hard [60]. Even the more narrow problem of tuning velocity profiles is NP-hard [3]. We consider the simpler, but more realistic scenario, given the UTM architecture, of scheduling UAS in real-time within lanes, reducing the configuration space of the UAS to a single dimension for each flight. The result is a practical, computationally tractable algorithm for strategic conflict management. The theoretical contribution here is an efficient algorithm for strategic deconfliction. We also provide experimental results that take into account the capacity constraints imposed by the system and enable airspace regulators to make informed decisions about how to address user demand.

The majority of motion planning algorithms relies on some form of discretization, e.g., cell decomposition or probabilistic sampling such as Rapidly Exploring Random Trees (RRT) [25, 60]. The algorithms that do not rely on discretization either assume a functional representation of trajectory (e.g., a spline) or are tactical because they apply to controls directly. The decisions related to discretization are vital in determining the effectiveness and complexity of a motion planning problem. For instance, in the RRT algorithm, the line connecting sampled locations must be discretely sampled to determine if any conflicts exist. If the sample resolution is too fine, then computation resources suffer. If the sample resolution is too coarse, then there is the possibility that a conflict exists that would not be discovered until it was too late.

The strategic conflict management problem shares characteristics with many application areas, as well as theoretical work in discrete mathematics (see [38] in the context of scheduling) and topology (see the chapter on configuration spaces in [25]). This includes the *Air Traffic Flow Management Problem* (TFMP) [14, 73, 83], *The Job-Shop Scheduling Problem* [7, 26, 54, 73, 75], *The Multi-Robot Motion Planning Problem* [84, 102, 119], *The Traffic Assignment Problem* [22, 74], and *Optimization Problems* [69, 121]. The FAA expects tens of thousands of UAS to utilize the airspace in close proximity over urban areas; therefore, the problem model composition is important to ensure that safety requirements are met. There are two ways in general to represent the safety requirements: using constraints, or with an objective function. The objective is to maximize the separation (or headway) between UAS. Assuming the solution is optimal, the question of whether it meets the safety requirement is determined by a threshold, e.g., "the minimum separation is at least 10 m," or "the minimum separation is at least 10 m with 99.9% probability."

We only consider the constraint model that casts the objective as a function of the time between desired release times and scheduled release times.

Lanes, as we propose them, are created by an authorized party (e.g., the Department of Transportation) and may allow USS proposed lanes that can be approved by the UTM authorities. Each lane has an entry point and an exit point and allows one-way travel from entry to exit. Where lanes intersect, we introduce an airspace structure inspired by roadway roundabouts. In addition, we provide a computationally tractable trajectory scheduling algorithm for UAS Service Suppliers (USS) within this structure. A capacity analysis follows the description of the airway structure to provide a baseline for further research. Prior research into the capacity of airspaces does not simultaneously consider the complexity of planning the operations; however, both concepts must be considered together since the airspace regulator is expected to manage both. We analyze the relationship between airspace capacity and such a lane-based structure. Over dense urban areas that are of primary concern here, there will most likely be a limited set of lanes possible, and understanding the capacity of the lane system is important to urban planners.

The lane-based method proposed here can be seen as an extension of Victor and Jet Airways used in manned air traffic management [34]. However, these were rigidly defined off of VOR systems (Very High Frequency Omnidirectional Range) in the 1960s. Moreover, such routes were under visual flight rules and at intersections required human deconfliction. The innovation in our approach is the dynamic nature of lane creation and deletion, as well as the introduction of roundabouts that permit efficient strategic deconfliction. Finally, we note that in the following, it is not assumed that UASs have the same speed in a lane, multiple levels (altitudes) are used in the lanes, and although lanes may be above roadways, this is not required; on the latter subject, it should be noted that NASA has stated (emphasis added) [112]:

> With regard to the routes that UAM will traverse between two vertiports, a *natural starting point* for emergent UAM operations is to fly along defined helicopter routes ... These helicopter routes tend to overlay highways and freeways on the ground to mitigate societal concerns

In the experiments described here, shortest route lane sequences are generated over urban areas (although arbitrary lane sequences may be used), but these are not large distance interstate routes, and the altitudes of lanes are somewhat arbitrary but are safely separated.

4.3 Lane-Based Strategic Deconfliction

Given a set of lanes created as described in Chap. 3, then UAS Service Suppliers (USS) will receive requests from UAS operators who want to obtain authorization to travel from a specific launch location through the lane network to a specific land

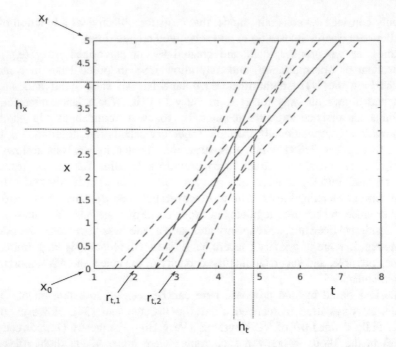

Fig. 4.3 Space–time lane diagram (STLD) for two UASs in a lane. The abscissa is time, and the ordinate is distance along the lane. h_t is the time headway (distance between UASs in time in lane), and h_x is the space headway (distance between UASs in lane). Note that h_t and h_x are linearly related due to the constant speed. The two trajectories in this scenario intersect at $t = 4$ and $x = 2$; however, they violate space headway before then

location. The launch and land locations will be the entry and exit points of the launch and land lanes, respectively. An approved flight plan for the UAS will consist of a sequence of lanes and corresponding lane entry and exit times. Given a sequence, $\mathcal{L} = \{\mathcal{L}_1, \mathcal{L}_2, \ldots, \mathcal{L}_n\}$, then the exit point of $\mathcal{L}_i$ is the entry point of $\mathcal{L}_{i+1}$, and the exit time from lane $\mathcal{L}_i$ is the entry time to lane $\mathcal{L}_{i+1}$. $\mathcal{L}$ must be specified as part of the UAS request; however, it is possible that UAS Traffic Management (UTM) systems may handle lane sequence selection differently.

In order to understand the interaction of flights in a lane, we have proposed the Space–Time Lane Diagram (STLD); this is similar to those used for ground road networks. Figure 4.3 shows a simple example of an STLD with two flights indicated by sloped line segments. The x-axis is for time, while the y-axis gives the distance along the lane. We assume that flights through the lane can be represented by their average speed through the lane. Thus, a flight, f_i, through the lane is indicated by a line segment with endpoint1 at $(t_{i1}, 0)$, and endpoint 2 at $(t_{i,2}, d_k)$, where $t_{i,1}$ is the time flight f_i enters the lane, $t_{i,2}$ is the time it exits the lane, and d_k is the length of lane k.

The other main issue is the determination of whether a proposed flight conflicts with any scheduled flight. An airway lane constrains the trajectory of the UAS to the center line of the airway, referred to as the longitudinal direction of the

aircraft trajectory in prior research (e.g., [52]). The vertical and lateral directions
are assumed to be under control to remain inside the lane. Uncertain altitude and
lateral movements should be compensated for in the design of the width and height
of the airway; this is a subject of ongoing research. Also, a constant velocity
is assumed within a segment; this constraint will also be relaxed in subsequent
research. Suppose that there exists a set of scheduled flights that are represented
in terms of lane enter–exit times and speed through each lane (the speed of a UAS
is assumed constant along a lane, but speeds may differ across UASs).

The Label Method
Let $F(c)$ be the set of scheduled flights through lane c defined as

$$F(c) \equiv \{r_{t_1}^1, r_{t_2}^1, s_g^1; \ldots; r_{t_1}^n, r_{t_2}^n, s_g^n\}$$

where $r_{t_1}^i$ is the lane entry time for flight i and $r_{t_2}^i$ is the lane exit time, and s_g^i is
the speed of the flight through the lane. Furthermore, let a flight request interval be
specified as

$$R \equiv [q_1, q_2, s_g^r]$$

where q_1 is the first possible launch time, q_2 is the latest possible launch time, and
s^r is the proposed speed. What must be determined is the set of (possibly disjoint)
intervals in R that are possible launch times (i.e., strategically deconflicted). In order
to determine this, the requested launch time interval is put in the Lane 1 STLD as
shown in Fig. 4.4a, where d_1 is the length of Lane 1, q_3 is $q_2 + \frac{d_1}{s_g^r}$, and q_4 is $q_1 + \frac{d_1}{s_g^r}$.
Each flight in Lane 1 is considered separately to ensure that the time headway, h_t,
is respected.

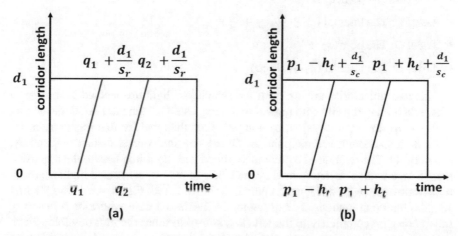

Fig. 4.4 Space–time lane diagrams: (**a**) trajectory boundaries for requested launch time interval
$[q_1, q_2]$. (**b**) The headway boundary trajectories for a scheduled flight that enters the lane at time
p and exits at time p'

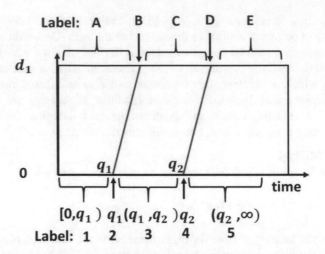

Fig. 4.5 Space–Time Lane Diagram Labels. 1,2,3,4,5 indicate intervals and times at the entry to the lane, and A,B,C,D,E indicate times at the lane exit

To determine safe launch time intervals, first consider the labeling of the STLD shown in Fig. 4.5. The labels are defined as follows:

- *Label 1*: The interval $[0, q_1)$
- *Label 2*: The point q_1
- *Label 3*: The interval (q_1, q_2)
- *Label 4*: The point q_2
- *Label 5*: The interval (q_2, ∞)
- *Label A*: The interval $[0, q_1 + \frac{d_1}{s_g^r})$
- *Label B*: The point $q_1 + \frac{d_1}{s_g^r}$
- *Label C*: The interval $(q_1 + \frac{d_1}{s_g^r}, q_2 + \frac{d_1}{s_g^r})$
- *Label D*: The point $q_2 + \frac{d_1}{s_g^r}$
- *Label E*: The interval $(q_2 + \frac{d_1}{s_g^r}, \infty)$

The two trajectories arising from the scheduled flight are labeled according to where their endpoints lie with respect to the requested launch interval. For example, if $p_{i,2} < q_1$ and $r_{t_1}^i + d_1/s_g^i < q_1 + d_1/s_g^r$, then the label for that trajectory is 1A since both start point and endpoint are in the first intervals at distances 0 and d_1, respectively. The relation of a previously scheduled flight in a lane to the requested launch time interval is determined by the labels of the two scheduled flight headway trajectories; the requested launch interval is shown in red. Figure 4.6 shows the first 13 possible combinations. For example, 1A,1A is the case where both headway trajectories are completely to the left (i.e., before in time) the first possible launch time trajectory through the lane. Note that in the figures, p_1 is $r_{t_1}^i - h_t$, p_2 is $r_{t_1}^i + h_t$,

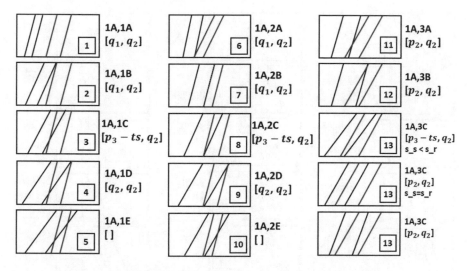

Fig. 4.6 Space–time lane diagrams for the first 13 possible label combinations

p_3 is $r_{t_2}^i + \frac{d_1}{s_g^i}$, and p_4 is $r_{t_1}^i + \frac{d_1}{s_i^e}$, and ts is the time for the requested flight to cross the lane. Also, the square brackets ([]) in the figure indicate the empty interval. Although there are 625 total label combinations, only 139 are physically possible; for example, no start time can be greater than the end time (see Appendix A for a complete enumeration and Appendix B for a Matlab program that determines the correct combination and returns the correct interval set). For each combination, it is possible to give the safe launch intervals contained in the requested interval (see the figure for some examples). In some cases, there is no possible safe launch time (e.g., 1A,1E in the figure). For other combinations, the resulting safe intervals depend on the relative speeds of the two UASs. An example of this is 1A,3C where a scheduled flight slower than the requested flight has a different interval as when the scheduled flight is equal or greater in speed. It can also happen that multiple intervals result as shown by the 2B,3C case in Fig. 4.7. To determine the viability of a flight through the complete sequence of lanes, each lane is considered in order as described by the Label Method Algorithm.

Algorithm 1: Label method

On input:
　lanes: lane sequence for requested flight
　$[q_1, q_2]$: requested launch interval
　n_c: number of lanes
　flights: flights per lane
　h_t: maximum required headway time

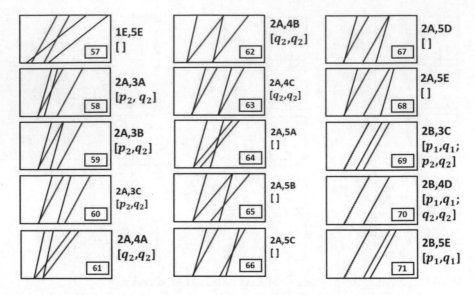

Fig. 4.7 Space–time lane diagrams for possible label combinations 57 through 71

On output:
    Safe time intervals to launch
begin
possible_intervals ← $[q_1, q_2]$
for each lane $c \in$ lanes
    time_offset ← time to get to lane c
    possible_intervals ← possible_intervals + time_offset
    for each flight, f, in lane c
        new_intervals ← $\emptyset$
        for each interval in possible_intervals
            $[t_1, t_2]$ ← interval i
            label ← get_label($r_{t_1}^f, r_{t_2}^f, s_g^f, t_1, t_2, s_g^r, h_t$)
            f_interval ← get_interval(label,$r_{t_1}^f, r_{t_2}^f, s_g^f, t_1, t_2, s_g^r, h_t$)
            new_intervals ← merge(new_intervals,f_intervals)
        end
    end
    possible_intervals ← new_intervals
end
possible_intervals ← possible_intervals - time to last lane

The key computational cost of this algorithm is the determination of f_int; each instance of this can be done in constant time; call it operation $\mathcal{I}$. Then given n lanes,

f_k flights in lane k, and f is the total number of flights in the lane sequence, then the total number of $\mathcal{I}$ operations is less than or equal to

$$\sum_{k=1}^{n} f_k + \sum_{i \neq j} f_i f_j$$

The second sum dominates the complexity, and assuming f_k is on average $\frac{f}{n}$, and since there are $\binom{n}{2}$ terms, then the big O complexity is $O(f^2)$.

An important point is that the lane-based formulation can apply in all cases where the trajectory is reduced to a single dimension. It is not required that lanes follow the road network on the ground; lanes may be organically created by operators and reused by other operators. The critical aspect of this formulation is that there are no crossing conflicts.

In general, we assume that the headway distance, h_d, is a known quantity. However, there are some constraints on how it is determined. Note that in the scheduling problem, the headway distance is converted into a headway time. The relation between the two is given by

$$h_t = h_d/s$$

where h_t is headway time and s is the speed of the UAS under consideration. The value for h_d is further complicated by the angle formed by two merging lanes (see Fig. 4.8). If UAS$_1$ is moving along lane i and UAS$_2$ along lane j, and they are to stay at least h_d apart, then there is a distance from the merge point, Q, of the two lanes that the other craft should remain outside. This distance, d, is given by

Fig. 4.8 Required distance along lane to avoid violating headway constraint

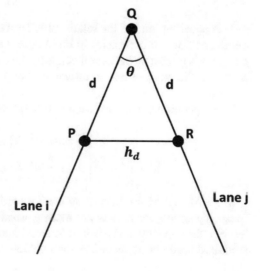

$$d = \frac{h_d}{2sin(\frac{\theta}{2})}$$

The UAS speed, s, to use is the slower of the speeds in its current or previous lane.

4.3.1   The Backprojection Method

The kinematic motion of the UASs may be described as follows:

$$uas_1 : x^1 = s_g^1\left(t^1 - r_{t_1}^1\right) \tag{4.5}$$

$$uas_2 : x^2 = s_g^2\left(t^2 - r_{t_1}^2\right) \tag{4.6}$$

where x^y is the longitudinal position (meters) within an airway segment for uas_y, s_g^y is the ground speed (meters/second), t^y is the time along the segment (seconds), and r_t^y is the release time, i.e., the time at which the UAS begins its trajectory across the segment. Also note from Fig. 4.4 that x_0 and x_f represent the start and end of the segment, so that $x_f - x_0 = length(segment)$. The time headway (distance between UASs in time) and the space headway (distance between UASs in space, sometimes referred to as *spacing*) are given by h_t and h_x, respectively.

The error bars in Fig. 4.4 represent the required spacing between UASs, also known as *well-clear* in the UAS literature. The vertical distance from a point on the line to the error bar is the *well-clear* and is denoted h_x^y for uas_y. Due to the linear nature of the problem, h_t^y and h_x^y are related by

$$h_t^y = \frac{h_x^y}{s_{g,y}} \tag{4.7}$$

This equation mirrors the relationship between density (or occupancy in space), speed, and flow (or occupancy in time) described in the *Highway Capacity Manual* [68]. This is important because it connects the concepts of road capacity, well known in road-traffic engineering, to airway capacity, which is explored in the following sections.

The separation constraints for any two UASs may be described as follows:

$$h_x = |x^1 - x^2| > max(h_x^1, h_x^2), \forall x^y : x^y \in [x_0^y, x_f^y] \tag{4.8}$$

$$h_t = |t^1 - t^2| > max(h_t^1, h_t^2), \forall t \tag{4.9}$$

Since h_t^y and h_x^y are linearly related, it suffices to consider only one constraint. These separation constraints are more general than the one considered in [52] to describe the capacity analysis in a foundational way. UAS operators may prescribe a required headway as needed by their vehicle and other operational considerations.

Consider the case where uas_2 is already scheduled and now a USS is presented with uas_1 to schedule. Since v_g^1 is considered constant, r_t^1 (the release time for uas_1) is the only decision variable. Let $h_{t,max} = max(h_x^1, h_x^2)$ and $r_t^1 < r_t^2$; we can describe the first position at which *well-clear* is violated by the following equation:

$$x_v\left(s_g^1 - s_g^2\right) + s_g^1 s_g^2\left(r_{t_1}^2 - r_{t_1}^1 - h_{t,max}\right) = 0 \qquad (4.10)$$

where x_v is the position along the segment where a violation first occurs. When the velocities are equal, then this equation reduces to the simple relationship,

$$r_{t_1}^1 = r_{t_1}^2 - h_{t,max} \qquad (4.11)$$

The corresponding constraint for planning purposes is then

$$r_{t_1}^1 < r_{t_1}^2 - h_{t,max} \qquad (4.12)$$

This assumption of uniformity of velocities is assumed in the experimental section to make network capacity constraints more visible. In the general case, however, when $s_g^1 > s_g^2$, then x_v is negative for all $r_{t_1}^1 < r_{t_1}^2$, and therefore, the only constraint is the same as Eq. 4.12. When $s_g^1 < s_g^2$, then the violation point may lie within the segment (this is the case in Fig. 4.3). The constraint is therefore

$$r_{t_1}^1 < r_{t_1}^2 - h_{t,max} - \frac{x_f}{m}, \quad m = \frac{-s_g^1 s_g^2}{s_g^1 - s_g^2} \qquad (4.13)$$

4.3.2 Backprojection Algorithm

The algorithm based on backprojection given here is a greedy scheduler (Algorithm 2):

Algorithm 2: Backprojection method

Require: $r_d, r_e, r_l, path, s_g$
    $r_d \leftarrow$ desired release time
    $r_e \leftarrow$ earliest release time
    $r_l \leftarrow$ latest release time
    $path \leftarrow$ requested segment ids
    $s_g \leftarrow$ speed
    $seats \leftarrow$ available time slots
    $l_s \leftarrow 0$ {The segment length}
    for each $segment$ in $path$ **do**
        $seats_{segment} \leftarrow$ seats on segment at $t \in [r_e, r_l] + \frac{l_s}{s_g}$

    $seats \leftarrow seats_{segment} \mid seats$ {Binary OR}

    $l_s \leftarrow$ segment length

end for

$r_t \leftarrow$ open seat closest to r_d

return r_t

It is called "greedy" because the scheduler only considers the currently requested operation and minimizes the distance between the scheduled and desired release times. In other words, it is locally optimal with respect to the desired release time. It is not globally optimal, in the sense that there may have been a better solution if all operations were considered simultaneously. In the UTM system, where operations are scheduled online and desired release times are unknown to the scheduler until the request is made, a globally optimal algorithm may not exist. To see why, this problem may be cast in terms of what Pinedo would describe as an online job-shop scheduling problem with no-wait constraints [75]. Specifically, this is an online-over-time problem because the scheduler "does not know at any point in time during the process how many more jobs are going to be released in the future and what their release dates are going to be" [75]. It is also classified as *clairvoyent* because all relevant information, such as speed, are available to the scheduler. It may be possible that a USS knows when its operations will be requested; however, it is still true that it will not know when another USS's operations will be requested (at least not in the currently envisioned UTM system). The no-wait constraint refers to the fact that, in the scenarios considered in this system, UASs cannot wait (park or hover) between successive segments. The problem of minimizing maximum lateness (a measure of the worst violation of due-dates), for a single machine with requested release dates (in Pinedo's nomenclature $1|r_j|L_{max}$), is NP-hard [75]. A polynomial-time online algorithm therefore represents an approximation of the optimal algorithm.

This algorithm applies equally well to homogeneous and heterogeneous velocities; however, only the homogeneous setup is considered here. The heterogeneous version of this algorithm applies additional time headway as required by the term $\frac{x_f}{m}$ in Eq. 4.13.

4.3.3   FAA-NASA vs. Lane-Based SD Comparison

Given the FNSD and LBSD approaches, the goal is to perform simulation experiments to better understand their respective advantages and disadvantages. To achieve this, some performance measures are defined. In addition to providing comparison metrics for the two methods, a set of measures for evaluation of different lane-based strategies is also described. Both of these are studied in terms of the following framework. A 100×100 unit area is considered, where 1 unit corresponds roughly to 10 feet. For the lane-based system, a 6×6 grid of ground locations is defined, and the subsequent airways based on that; nearest eight neighbors are connected, and every ground vertex has both launch and land lanes. For the FAA-

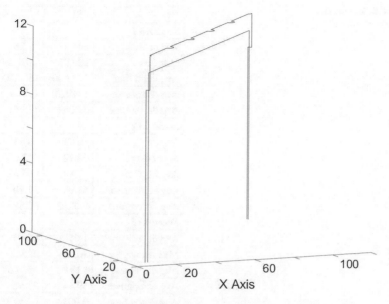

Fig. 4.9 The lane-based (blue) and FAA-NASA (red) routes from ground vertex 1–36

NASA flights, a 5×5 grid is defined (i.e., grid elements are 20×20 square units). UAS speeds are in the interval [0.1,0.31], as these correspond to 3–10 mph. The altitude for the lanes is between 10 and 12 units, while for the FAA-NASA flights, it is set to 11 units. All flights are specified as between two ground vertexes, and lane-based flights take a shortest route through the lanes, while the FAA-NASA flights follow a 3-polyline trajectory of up, over, and down. Note that this makes these latter routes shorter than the lane-based routes. A set of 1000 flights is scheduled in each scenario; however, if the deconfliction takes more than 30 seconds for some flight, then data from the first 75 flights is used to interpolate a result for all 1000 flights.

Within this context, we consider three scenarios:

- *Scenario 1*: The launch and land ground vertexes are selected to be the two most distant (i.e., 1 and 36). Figure 4.9 shows the lane-based and direct routes for this. The speed of every flight is fixed to be 0.12, and all flights follow the same trajectory. The start times interval for these flights is [0,2000]; that is, a flight should be assigned the earliest possible launch time in this interval. Finally, the minimum headway distance is set at 1 unit.
- *Scenario 2*: This is the same as Scenario 1, except that the launch and land ground vertexes are chosen randomly.
- *Scenario 3*: In this scenario, the launch and land ground vertexes are chosen randomly, as are the UAS speeds, and the start times interval. Each UAS has its own speed that is constant across the whole flight, and the initial start time is randomly selected in the interval [0,1000].

Table 4.1 Results of
experiments

	LSD	FAA-NASA
Scenario 1		
Avg delay	494.97	583.61
Max delay	989.95	1167.20
Avg flight time	1527.00	1318.00
Avg comparisons	1249.50	1.49×10^7
Avg decon time	0.0063	138.53
Scenario 2		
Avg delay	19.87	10.73
Max delay	67.72	917.60
Avg flight time	879.49	705.68
Avg comparisons	61.20	27.40
Avg decon time	0.0212	0.2524
Scenario 3		
Avg delay	1.65	16.85
Max delay	30.39	325.20
Avg flight time	532.88	451.65
Avg comparisons	49.84	28.62
Avg decon time	0.0014	0.3579

Note that Scenario 3 is closest to a real situation.

Table 4.1 gives the data collected from the experiments. As can be seen, the lane-based method does better in Scenarios 1 and 3, while the FAA-NASA approach performs better on average delay, but not maximum delay, on Scenario 2. Scenario 1 represents the scheduling problem on a heavily used route, while Scenario 3 is more representative of a random arrival process; thus, we believe that these are more reflective of actual operational situations. Another observation is that these results are achieved in the context of all flights being nominal, that is, no contingencies occur. Since the lane-based approach has a distinct advantage with respect to contingencies, then these results indicate the overall superiority of the lane-based approach. Also, note that cost of deconfliction in the lane-based approach in Scenarios 1 and 3 is two to four orders of magnitude lower.

Another advantage of the lane-based method is that it is possible to easily visualize the flight schedules through the lanes. Let us consider an example from each scenario. First, consider flight 10 in Scenario 1. Figure 4.10 shows the complete lane sequence for the flight with all other scheduled flights. Flight 10 is shown in red, and since all the flights follow the same lane sequence and have the same speed, they are all represented as parallel line segments where the lower endpoint represents the launch time, and the upper endpoint represents the landing time. It is also clear that this representation makes it easy for a flight operations center controller to visually determine if flights are off course by overlaying telemetry data on top of this graph. Figure 4.11 shows the corresponding Space–Time Lane Diagram for Scenario 2. Here it can be seen that Flight 10 is the only flight scheduled along this specific route, but that other flights are scheduled at various times on some of the lanes in the route. Finally, Fig. 4.12 demonstrates how readily system-wide type information

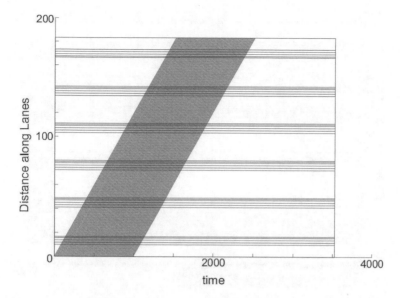

Fig. 4.10 The stacked set of space–time lane diagrams for flight 10 in scenario 1

is made evident by these graphs; note that the number of segments in the upper lanes (later part of the flight) indicates that there may be some congestion in that region, and it might be wise to find alternate routes so as to avoid that. The variety of slopes in the graph indicates the different speeds of the flights through the lanes.

4.3.4   Lane Stream Properties

We now define properties specific to the lane-based approach. To do so, we assume an airway lane of length d and consider a time interval of length t_{max}, call it $[0, t_{max}]$. Also assume that all UASs fly through the lane with a constant speed, s. A flight scheduler assigns start times for flights to go through the lane; let S be a set of such start times. Then, to satisfy constraints, it must be the case that no two start times are closer than headway time, h_t, of each other. This is equivalent to packing segments of length h_t into the lane (time) interval. Note that $h_x = s \cdot h_t$ is the headway distance. The maximum number of UASs possible in the lane at one time, n^t_{max}, is then

$$n^t_{max} \equiv \lfloor \frac{d}{s \cdot h_t} \rfloor + 1$$

Clearly, achieving n^t_{max} depends on obtaining a perfectly packed requested start time sequence.

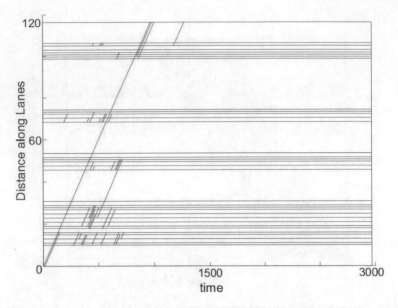

Fig. 4.11 The stacked set of space–time lane diagrams for flight 10 in scenario 2

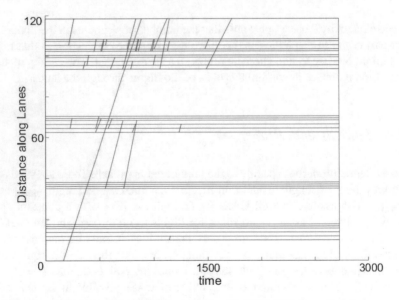

Fig. 4.12 The stacked set of space–time lane diagrams for flight 10 in scenario 3

Suppose that flight request start times are sampled from a uniform distribution across the given time interval $[0, t_{max}]$. The *time occupancy*, $\Theta_t(\mathcal{A})$, is a function of the scheduling algorithm $\mathcal{A}$ and is defined as

$$\Theta_t(\mathcal{A}) \equiv \frac{\mu_{\mathcal{A}}}{n_{max}^t}$$

where $\mu_{\mathcal{A}}$ is the mean number of flights through the lane during the time interval $[0, \frac{d}{s}]$ of several trials with algorithm $\mathcal{A}$. If the scheduler has no choice but to assign the requested start time if possible and otherwise reject the request (call this algorithm $\mathcal{A}_0$), then this is an example of Renyi's Parking Problem [27, 39, 80], and $\Theta(\mathcal{A}_0) \to 0.74759$ as $t_{max} \to \infty$. In the experiments below, we compare algorithms and lane parameter sets by means of their observed time and space occupancy measures.

Next consider standard ground traffic stream properties: density, occupancy, and flow (see [114] for a detailed discussion). The *spatial density* of the lane at time t, $k_s(t)$ is defined as

$$k_s(t, \mathcal{A}) \equiv \frac{\mu_{\mathcal{A}}}{d} \tag{4.14}$$

that is, the average number of vehicles in the lane over the length of the lane. *Spatial occupancy* can then be defined as

$$\Theta_s(t, \mathcal{A}) \equiv \frac{\Theta_t(\mathcal{A}) \cdot n_{max}^d}{d} \tag{4.15}$$

Finally, *spatial flow*, $q_s(t, \mathcal{A})$, is defined as

$$q_s(t, \mathcal{A}) = k_s(t, \mathcal{A}) \cdot s \tag{4.16}$$

These traffic stream properties are used to characterize the performance of a set of algorithms compared in the experimental section.

These measures are given as a means of comparing the effectiveness of alternative lane scheduling algorithms. Since that problem is not addressed here where we compare the FAA-NASA approach to lanes, we simply give the values for these measures for Scenario 1, where the flights are most densely packed. The following values result for Scenario 1 for the launch lane:

$$n_{max}^t = 8 \tag{4.17}$$

since $h_x = 1.41$

$$\Theta_t(\mathcal{A}) = \frac{8}{8} = 1 \tag{4.18}$$

looking at the time interval $[0, 83.333]$ (since $\mu_{\mathcal{A}} = 8$ and $83.333 = \frac{10}{0.12}$).

$$k_s(t, \mathcal{A}) = \frac{8}{10} = 0.8 \tag{4.19}$$

$$\Theta_s(t, \mathcal{A}) = \frac{1 \cdot 8}{10} = 0.8 \tag{4.20}$$

$$q_s(t, \mathcal{A}) = 0.8 \cdot 0.12 = 0.096 \tag{4.21}$$

A direct comparison of performance characteristics has been made between the FAA-NASA and lane-based UAM approaches. A variety of scenarios were examined, and measures defined on the computational and other requirements over a set of flights. The lane-based method was found to outperform the FAA-NASA approach in the most likely actual conditions that will be encountered in large-scale UAS traffic management. Although the lane-based method requires flights of slightly longer route, there are multiple advantages in terms of management. Finally, the incorporation of manned drones in this system is possible so long as the human pilots follow the assigned flight path; in addition, human pilots would be in a better position to handle contingencies.

In Chap. 8, we describe the use of Agent Based Modeling and Simulation (ABMS) to determine more optimal UAM parameters related to lane properties and their layout, as well as lane speeds, and auxiliary lane support structures (e.g., emergency lanes alongside regular lanes, emergency landing lanes). In addition, Chapter 6 looks into real-time adaptive lane scheduling by the UAS themselves. This may be particularly useful locally in contingency situations. We are exploring the formal verification of the safety aspects of such protocols. Finally, we are working with the Utah Department of Transportation to realize a version of lane-based UAS traffic management in urban regions (e.g., the Salt Lake City Valley) in order to effectively meet the challenge of large-scale UAS deployment for deliveries and other services.

Chapter 5
Air Traffic Operations Center

5.1 Introduction

The AAM community, including UAS service providers, operators, and relevant government authorities, aims to provide a wide number of services (e.g., package delivery, air taxi, etc.) by means of robust and safe UAS Traffic Management (UTM) systems that achieve large-scale (i.e., thousands per day) operations in urban areas without human control, but with reliable communications and contingency planning (see [82]).[1] UAS Service Suppliers, for example AirMap [2], have dealt with the operational interfaces, integration of Geographic Information Systems (GIS), registration of flights, UAS communications, and monitoring UAS activity. These capabilities are all developed to operate as described in the strategic deconfliction context defined by NASA [85] that is defined in terms of a geographic grid (a set of cells). Each new flight must be deconflicted pairwise in terms of grid cells that have other scheduled flights. The trajectory of a UAS flight is a curve in 4-dimensional space (x,y,z,t). Given a set of such curves, strategic deconfliction (i.e., make sure that no two flights are ever within a specified minimum distance called *spatial headway*) necessitates a pairwise comparison of the curves with motion constrains, and determining a good or optimal trajectory in this configuration space is in general PSPACE hard. Moreover, the FAA and NASA have yet to specify contingency protocols, and some research suggests that all flights simply return to base in these scenarios (e.g., lost-link).

In previous chapters, we have proposed and studied various aspects of a lane-based approach [89, 92, 97–100]. This lane-based approach reduces strategic deconfliction complexity (to 1D from 4D) and makes the handling of contingencies a spatially local problem [90]. The use of lanes for commercial flights (Victor and

---

[1] This chapter was co-authored by Vista Marston.

Fig. 5.1 Lane-based airway lanes above downtown San Francisco, CA

Jet Routes) has a long-standing history [12]. However, human air traffic controllers manage commercial airway lanes, and this management function that must be automated if a large number of autonomous flights are to take place daily over major metropolitan areas.

Given a lane-based UTM system, then lanes are created as a static structure (much like ground road networks), and all scheduled flights will follow a sequence of assigned (reserved) lanes from launch to landing. Figure 5.1 shows a lane-based airway over downtown San Francisco, CA.

If all UAS flights performed in the lane-based system as planned, and no unexpected events occurred (e.g., unplanned flights entering), then there would be no need to monitor the airspace. But since unexpected things do happen, some sort of traffic management center is necessary to help mitigate the effects of unforeseen events. In addition, an air traffic control operations center (ATOC) can serve to apply UTM policies and enforce the rules as necessary. ATOC operators can monitor activity in the lanes and use existing conditions as part of the flight plan approval process. For example, if there are strong winds in part of the airspace, the ATOC can provide advisories and route flights away from the impacted area. In addition, the ATOC can gather airway performance data and modify policies accordingly to improve things.

It is also possible that existing ground transportation infrastructure can be leveraged to support ATOC operations. For example, if road intersections already provide power (say for cameras at traffic lights) and network access, then upward looking radars can be placed there to observe the airways. It may also be useful to have close coördination of the ATOC and ground traffic management operations,

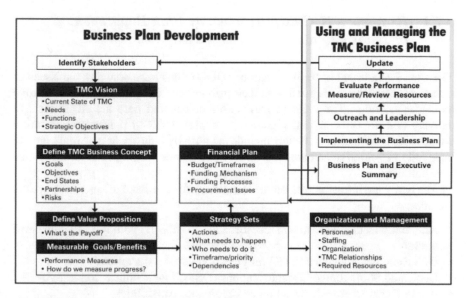

Fig. 5.2 ATOC business plan (from [20])

especially if UASs fly over roads. This allows rapid dissemination of disruptive events like an emergency landing on a road.

It is most likely that ATOCs will improve transportation performance. Certainly, this has been the case with ground transportation management systems (GTMS). For example, the TransStar system in Houston, Texas helped reduce traffic delay by over eleven million hours per year. In Utah, the CommuterLink system decreased intersection delay by 27% and increased freeway speed by 20% [110].

A full-blown GTMS for a large metropolitan area can be a costly proposition, and the cost for an ATOC would presumably be similar. A ball-park cost projection given by [20] is about $3M for a building, and another $3M annually for operation and maintenance. Of course, monitoring or provisioning hardware infrastructure imposes more cost. The same study gives a Business Plan Development Guide that shows the steps necessary to develop a Traffic Management Business Plan (see Fig. 5.2). Planning for an ATOC would have a similar set of considerations. However, one major additional requirement would be for some form of interaction with both the FAA and also directly with UAS flights. Working with the FAA is necessary to maintain the safety of simultaneous manned and unmanned air activity. Communication with UAS operators is necessary to receive real-time telemetry data from UAS flights.

5.1.1 Example ATOC Requirements by Utah Department of Transportation

The Utah Department of Transportation[2] (UDOT) has been studying the issues of Urban Air Mobility since 2018 with their prime focus on the airspace of the Provo to Ogden corridor. The main two use-cases considered here are small package delivery and air taxis. A major goal is to exploit UAS to reduce ground traffic congestion. Other pertinent factors include the push by industry (e.g., Airbus, etc.) to have full integration of autonomous vertical takeoff and landing vehicles for human transportation.

The hardware infrastructure deemed necessary for a reliable urban airway system includes:

- Automatic Dependent Surveillance-Broadcast (ADS-B) with links to unmanned aircraft
- GPS Real-Time Kinetics (RTK) for precision vertical operations
- Dedicated Short-Range Communications (DSRC) for seamless connectivity
- Active Radar to provide detection and avoidance capabilities

Figure 5.3 shows the radar and other coverage proposed by UDOT. Figure 5.4 shows the GPS coverage network and example hardware. Figure 5.5 shows the current ATMS and fiber network capable of supporting UAM in the main Salt Lake Valley corridor. Finally, UDOT proposed the acquisition of (1) several more GPS-RTK reference stations for safety redundancy, (2) Mini-radar/ADS-B with an estimated 40 stations needed covering about $1267 \, km^2$ with power and network connections already available through UDOT, and (3) DSRC wireless. The estimated cost for this infrastructure system was estimated as about \$200K for GPS reference stations and \$12.7M for mini-radar/ADS-B hardware.

In terms of ATOC UTM capabilities, UDOT established goals to:

- Manage low-altitude air traffic
- Provide real-time notification of flight plan changes
- Communicate hazards or priority routing
- Account for traffic load, impact of skyport placement, weather, rogue, and emergency air traffic

UDOT also proposed that UAM traffic be routed over existing roads and along planned flight corridors with pre-approved flight plans, with assigned altitudes related to direction of flight and with speed and location constraints. Figure 5.6 shows the proposed air layout. Of course, complete UAM systems will most likely be built up incrementally as technology, demand, and resources converge to make it possible. Current UAS platforms require a great deal of improvement and demonstrated robustness in order to meet the stringent safety requirements of air

---

[2] We would like to thank Jared Esselman for this material.

Fig. 5.3 Radar and other sensing capabilities proposed by UDOT

Fig. 5.4 GPS coverage and example hardware

transportation. UAS management software is not yet available, the package and air taxi industries are not yet at scale, and air monitoring and enforcement mechanisms have yet to be developed and demonstrated.

Thus, the development of Advanced Air Mobility (AAM) systems requires not only a robust and safe approach to planning flights, but also a way to monitor UAS flights in real time to determine whether flights are deviating from their

Fig. 5.5 ATMS and fiber network for Salt Lake Valley airway corridor

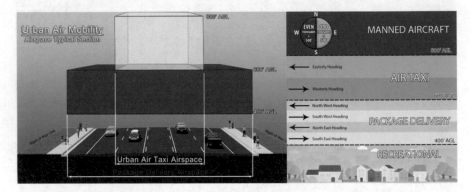

Fig. 5.6 Proposed airspace layout for the Salt Lake Valley airway corridor

nominal flight paths or if there are rogue (i.e., unplanned) flights in the area. We have proposed a lane-based airways methodology for lane creation, scheduling, and strategic deconfliction, and here we describe *Nominal vs. Anomalous Behavior* (NAB) in an efficient and effective way to monitor flight trajectories to determine normal versus anomalous behavior.

5.2 Lane-Based Monitoring

The basic problem addressed here is how a lane-based UTM system supports the recognition of rogue flights of a variety of sorts: amateur recreational hobbyists, UAS operators making an unscheduled up, over and down flight, malicious operators, etc. In order to detect such rogue flights, we propose the analysis of trajectories based on their deviation from the lane structure, including both location in space

Fig. 5.7 The set of airway lanes created over Salt Lake City, UT

and direction of flight at that location. The basic idea is that a model can be produced directly from the lane structure and compared to any flight individually. The alternative FAA approach would require knowledge of all 4-dimensional flight trajectories, as well as multi-target tracking to monitor the flights along those curves and a comparison of an unidentified flight to all of those curves. Thus, the proposed lane-based method is much more efficient and effective.

5.2.1   NAB Modeling

Given a set of UTM lanes, a convenient model is just a set of point samples on the lanes, each with an associated direction of the travel in the lane. Figure 5.7 shows a lane-based airway over the East Bench area of SLC, UT. Figure 5.8 shows a set of sample points from the lanes. These provide a good model since any nominal flight (i.e., following its assigned lane sequence) should be near a lane and headed in the direction of the (one-way) lane. As part of the model, the direction vectors can also provide significant information about a flight. Figure 5.9 shows a subset of the trajectory direction vectors used in the model.

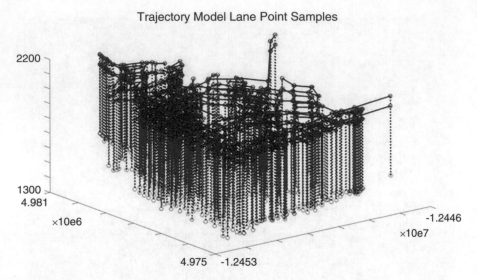

Fig. 5.8 Trajectory point set model of airway lanes over East Bench of Salt Lake City, UT. Red circles are lane endpoints; blue points are samples along lane

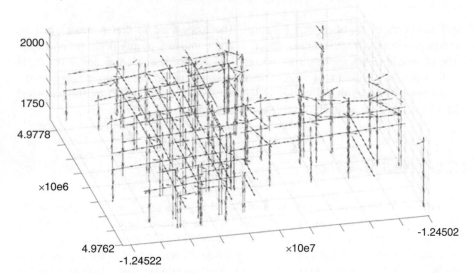

Fig. 5.9 Trajectory direction vector model of airway lanes over East Bench of Salt Lake City, UT

5.2.1.1   The Lane-Based Trajectory Model

A trajectory is a sequence of 4-tuples (x,y,z,t) that provides a representation of the actual UAS motion through space and time. Trajectories can be divided into two basic types: *expected* and *unexpected*. An expected trajectory is one that results from a planned flight trajectory perhaps combined with some error due to contextual influences (e.g., weather, power, etc.). An assigned trajectory in terms of the lane structure is a sequence of straight line segments, but when exiting one lane to enter another, the actual flight path deviates from the sharp corner turn. Wind and precipitation may also cause a flight to be off-course. To account for this, lanes are viewed as having an enclosing 3D volume (e.g., a tube structure) containing the straight line segment, and this allows room for error. In addition, the spatial headway constraint provides more cushion for speed changes, etc.

Trajectories may arise from various flights, but a scheduled UAS is required to provide telemetry data at a set update rate; this data comes from onboard sensors. Alternatively, radar or other sensors may be used to monitor flights and provide an independent source for trajectory data.

Next, consider a planned flight and its associated trajectory. When the UAS sends telemetry data, it also sends its ID. This makes it possible to determine if the flight is off-course and by how much. It is also possible to monitor the airspace and produce locations of airborne objects. We assume that it is possible to determine which objects are UAS with high probability (as opposed to birds, etc.). Given such data, it is possible to corroborate UAS telemetry location data. The result is that planned flights have expected trajectories in that they are near the planned path or their reported locations are consistent with ground sensor data.

On the other hand, a flight that has not been planned produces a trajectory that corresponds to the type of the flight, and, in general, the trajectory will not correspond to any planned flight, will not be corroborated by sensor data, and will not follow a sequence of connected lanes. This is an unexpected trajectory. Of course, there are ways that UAS can insert themselves into the lane structure and mimic a scheduled flight (e.g., by following a scheduled flight), but this can also be detected in that they do not provide telemetry data. Unscheduled flights in the airspace are called *rogue flights*, whereas unexpected trajectories are called *anomalous*. In practice, the ATOC needs to detect rogue flights as robustly as possible. The NAB method is one such approach.

NAB operates as described in Fig. 5.10. The lane data is made available along with the UTM policy parameters and the set of scheduled flights. Based on this, a spatial database is constructed consisting of a set of 3D points sampled along the lane and, to each of these points, is associated the direction of travel vector at each of those points (recall that lanes are one-way). The lane model consists of this data organized so as to be efficiently exploited. The inter-sample distance must be selected so to keep the number of points down while at the same time allowing adequate discriminatory power to determine if a flight is near a lane and headed the right direction.

For example, consider the Salt Lake City East Bench airways shown previously. If an inter-sample distance of 2 m is chosen, it produces a set of 454,331 points. In order to keep the computational complexity low, the points are organized as a kd-tree using the 3D points. A kd-tree divides the points at the median of each of the vector dimensions in turn (or picking the dimension with the greatest spread in values in

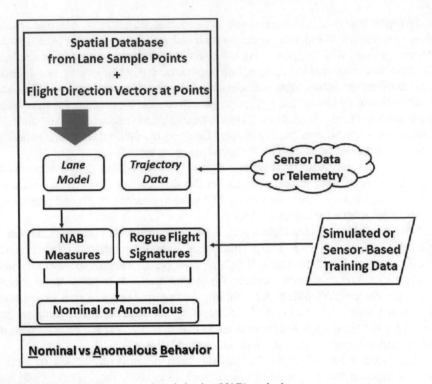

Fig. 5.10 The nominal vs. anomalous behavior (NAB) method

turn) and thus achieves a relatively well-balanced tree that allows $O(log(n))$ search time to find the closest points in the spatial data to a query point. Any nominal flight should be near one of the sample points and headed in the appropriate direction. Of course, a temporal analysis can be performed by checking the associated Space–Time Lane Diagram (see [92]) that specifies the position of each flight in a lane at each time instant. Also, with the FAA-NASA approach, there is no fixed set of lanes, and therefore, every existing flight would require target tracking against the set of all flights. Since this approach requires large-scale sensing and computing resources, it is not considered further here.

Next a set of NAB measures are determined, which allow the discrimination of the different types of flights, both nominal and rogue. These are computed either by comparing the UAS trajectory to the lane data, or simply in terms of the trajectory itself. For example, two lane related measures are:

1. M_{dist}: minimum distance to a lane at each time step
2. M_{dir}: cosine of the angle between the UAS direction of travel and the lane direction of travel at each point

These measures are applied at each point in the trajectory to produce a temporal signature to represent the flight. An example of a measure based solely on the

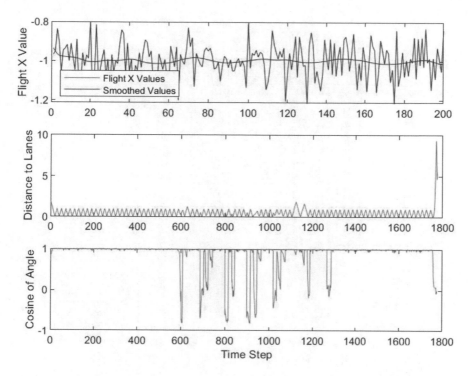

Fig. 5.11 NAB measures for a nominal flight

trajectory data would be the amount of time spent hovering (i.e., staying for some minimal duration in time in one place in space). Given a characterization of the types of flights of interest, then a set of trajectory signature templates can be constructed and used as class models. Such templates can be the result of a set of simulations or produced from datasets of actual flight trajectories. Given a new trajectory, its measured features are compared to the flight signature templates and matched to the closest in order to classify the type of trajectory (i.e., nominal or anomalous).

Consider a nominal flight that does not perfectly follow the lane but rather has some noise associated with it. Figure 5.11 top row shows the x values of a nominal flight trajectory (with a Gaussian noise of 0.16 variance) and a smoothed version of that data (in red). The middle row shows the distance to the closest lane, and the bottom row gives the cosine of the angle between the direction of flight and the lane direction. This distance and direction difference are NAB measures. For the distance measure, over 96% of the trajectory points are within 1 unit of the lane, and for the angle difference measure, 70% are within 10°. The large angle differences arise at lane changes.

Now consider rogue flights. Five categories are explored here:

- *Hobbyist Type I:* Flies up from one place and makes a few moves above the launch site and then eventually lands at the same site.

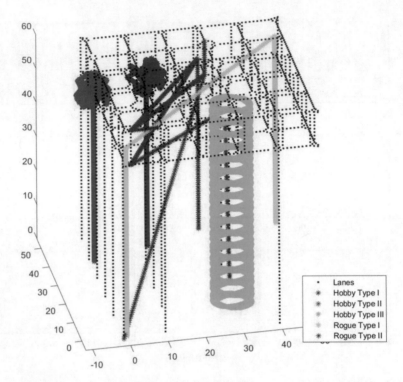

Fig. 5.12 Examples of the five anomalous flight trajectories

- *Hobbyist Type II:* Flies up from one place and makes a few moves above the launch site, hovers after each move, and then eventually lands at the launch site.
- *Hobbyist Type III:* Flies up in a circular motion to some highest point and then flies down in a circular motion to land.
- *Rogue Type I:* Flies up over and down as for a delivery.
- *Rogue Type II:* Flies up to a lane, flies along the lane to the end, then flies to another lane (not necessarily connected), and eventually flies down to land.

These anomalous flight patterns are representative of the types of flights to be expected. Figure 5.12 shows an example of these types.

5.2.1.2 Sensor Data to Track Rogue Flights

Radar is used to scan the airspace lane network so as to make sure that scheduled flights adhere to their flight plans and to detect unscheduled flights. Here we provide an overview of the considerations necessary to employ radar sensors in airspace monitoring. Although the description addresses radar in the context of the simulation system, the actual exploitation of radar systems would require a similar analysis.

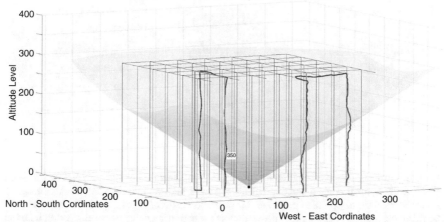

Fig. 5.13 Demonstrates how a vertical radar field would be placed in a grid layout lane system simulation

In order to analyze flights in the simulation, there first needs to be a monitoring system that can detect UAS flights and report flight information back to the Air Traffic Operations Center (ATOC). The monitoring system that was selected was Radar. The structure and performance of the radar's objects was to imitate real radar systems. One can think of the system as the radar sending signals to detect objects over certain time intervals. If an object is detected, it will be reflected through broadcasting to those that are watching the scanner. This section will discuss the mechanism used to achieve this performance and structure.

Figure 5.13 shows a single radar field in the grid layout lane-based simulation. One might notice that the radar's detection field is conical in shape as the field progresses through the atmosphere, and this is due to the natural occurrence of refraction. In real life, this radar field would not contain distinct lines as a result of attenuation, meaning that the radar's signals on the outer edges are weaker than those closer to the center. One can think of it similar to a flashlight shining in a dark room, where the center is the brightest and gets dimmer moving to the outer edges. However, in our simulation, we excluded these weaker areas and created distinct edges for simpler computations that are discussed later in the chapter.

Radar Placement

Radar systems can be selectively placed within city coordinates in order to provide the most coverage of a given airspace region that surrounds the lanes. There is a lot of variability in the creation of the radar objects that can affect the coverage (see Fig. 5.14). For example, some of these features are the maximum range of the

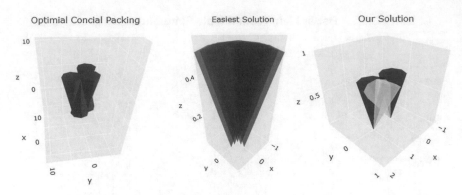

Fig. 5.14 Displays a couple of different packing arrangements for conical shapes. The image on the left is the optimal conical packing but not the most realistic solution. The middle image is the easiest solution to optimize coverage, but it is too costly of a solution to be implemented. The rightmost image is our solution, which provides a high percentage of coverage with minimal overlaps

radar field, the direction of the radar field, the beam angle, and the placement of the radar itself. Our goal is to find radar placements that provide the best coverage while minimizing the amount of radar field overlaps.

First let us make an assumption that the area we wish to cover is a cube that encapsulates the total area around the lane structure. The reason we chose to maximize the radar fields' coverage of a cube instead of the actual lane structure was because we wanted to ensure additional coverage beyond the lane system that would allow for the detection of rogue flights. In addition, turning the region of interest into a cube allows us to solve our problem by using packing optimizations for conical structures. The best solution would be to set the length of the radar's field to the length of the cube. Then alternating the radar field direction in the vertical direction while alternating which edge of the cube to place the radar on, which is shown in the figure below. However, this solution is not really practical for a real life scenario, because it has radar systems that are floating in the air. Unfortunately, it would be impractical to suspend radar systems in midair like this. A more practical solution is to mount the radar systems to other objects that are on the ground such as on buildings, light posts, or the ground itself. Therefore, the arrangement of radar systems must also take into account their physical mounting requirements. The easiest solution would be to place radar systems away from each other at small increments. This would provide significant coverage but would also come with a significant price tag if it were to be implemented in the real world.

Our solution takes a different approach; instead of placing several radar systems on top of one another, we spaced them based on the height and radius of the radar fields. No matter how many lanes are stacked on top of one another, we decided to use the lower lane height as the point where the radar fields are okay to overlap; this ensures that the entire lane system will be covered. To find the distance between

Fig. 5.15 The basic
components of a conical
shape; these are used to
establish the required distance
between radar placements

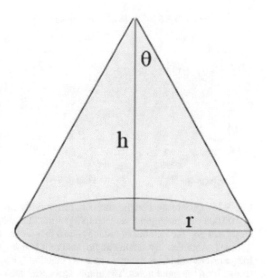

radar systems, we take the radius of the radar field at the lower lane's height and
then double this radius. The final result is the distance between the radar systems.
Figure 5.15 shows the basic parameters of a cone, and given that

$$tan(\theta) = \frac{radius}{height}$$

then we have

$$radius = height \times tan(\theta)$$

Let us shift the attention to how to measure the coverage of a set of radar systems.
Getting an exact measurement is quite difficult; therefore, an approximation of
coverage is calculated using Monte Carlo experiments. In these experiments,
randomly generated samples, which are bound to the cube area around the lane
system, will be used to count the number of points detected by the radar systems. To
approximate the volume of coverage, we can simply divide the number of detected
samples over the total samples in the experiment. It is through these experiments
that we can determine the differing factors that help maximize coverage. Figure 5.16
shows the effects of changing the beam angle through the Monte Carlo experiments.

Due to the conical shape of the radar fields, most of the coverage that is gained
when using this method is in the upper altitude lane structures; however, there are
some regions that are left uncovered, which happens to correspond to areas that
could contain launch and landing lanes, which must have coverage if this system is
to function safely in the real world. There are two potential solutions: first, placing
radar systems at every launch and land lane, or second, including more horizontally
directed systems.

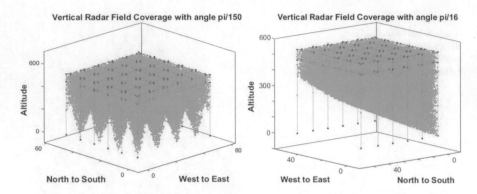

Fig. 5.16 Difference comparison between varying beam angles in the Monte Carlo experiment of 100,000 randomly generated sample points. This displays the inverse relationship between the angle size and the number of radar systems that would be needed for coverage. The image on the left uses a significantly smaller angle, resulting in more radars to cover the given area, whereas the image on the right uses a larger angle of detection. The green dots represent the sample points that were within the radar fields. The empty spaces are areas in the lane system that are not covered by the radar systems that have been placed

There will be a crossover point between placing radar systems at every launch and land sites and just placing horizontally directed radars. If there are a significant number of launch and landing sites relatively close to one another, there will also be more radar systems close to each other, leading to greater overlap between radar fields. In contrast, strategic placement of a few horizontally directed radar systems could cover a larger area while also reducing the amount of overlap between systems, meaning the number of radar systems needed will be lower to maintain the same amount of coverage. For example, consider the lane structure in the figure above with a radar angle of $\frac{\pi}{16}$, and it contains 6 launch and 6 land sites for a total of 12 sites that should be covered. Therefore, 12 additional radar systems would be needed to cover all of these sites using the first approach, whereas the second approach would only need one horizontal radar system to cover the remaining area, if it were placed in the same location as the other original radar.

Now consider how these radar systems detect objects during the simulation. When detecting objects, the main goal is to determine whether the object falls within at least one individual radar field. When determining if an object is within a radars field, one must consider a couple of factors: distance away from the radar, the relative angle from the radar, and the radar's maximum range.

To simulate a more realistic scenario, every radar and object have to be looped over to check which object falls into which radar field. Therefore, the time complexity of this check would be $O(RO)$, where R is the number of radar systems and O is the number of objects in flight during a time step. Therefore, since there could be a significant amount of radar systems and objects to detect, the calculation must be simple enough to ensure that the time complexity does not become excessive. To keep the time complexity manageable, there are only two

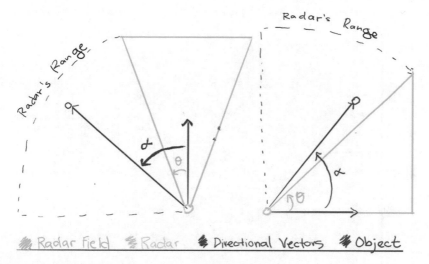

Fig. 5.17 Properties that are used to determine if an object is within the radar's field of view. These properties include the radar's maximum range, and the angle between the object and the radar. The left drawing shows a vertical radar field, and the right drawing shows a more horizontal radar field

checks involved when determining if a given object is within an individual radar's field. The first check is to see if the object is within the radar's range, and this is done by calculating the Euclidean distance from the object itself to the radar. The second check is to find the relative angle between the object and the middle of the radar field (Fig. 5.17). For this, two directional vectors are created: one to represent the radar field itself using the middle of the beam, and the other to represent the directional vector between the object and the radar itself. Then using the dot product to find the angle between the two directional vectors allows for the comparison between this angle and the radar beam angle.

During each simulation step, each radar object will receive a list of objects that are in flight. Then each radar will perform a detection step for each individual object. Each radar object contains a list that is clear at the beginning of each simulation step and then replenished with all of the objects that were detected. Once the radar object has gone through all of the objects in flight, it will then broadcast to any listeners that it has detected objects in the given simulation step. Any subscribers that are listening will have access to this list of objects, and they can proceed to handle the information according to their guidelines.

Each simulation will contain an individual lane system, radar system, the number of flights, and an individual ATOC. As the simulation is running, information between all of these components is hard to capture until the entire simulation is completed. Therefore, each of these classes contains functions that allow you to gain some insights, while the simulation is running in order to prevent a complete black box of the simulation process. The radar systems method is to display the

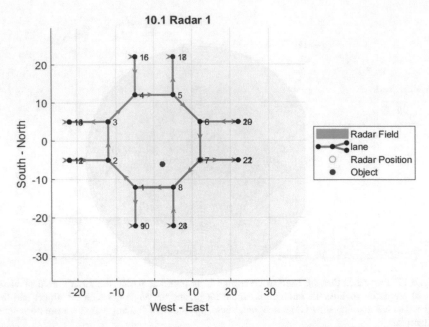

Fig. 5.18 Visualization of the radar detection process during a simulation. The colorful circle represents the radar field, and the red dot represents the location of the detected object within the field. The first number in the title represents the time step and is followed by the radar identification number

region it covers, along with a red dot to represent the location of any detected object at that given simulation step. Figure 5.18 above displays the graph.

This is a useful tool for many reasons, one being able to target locations that might contain a higher density, or an area that potentially contains a lot of anomaly flight behaviors. The user can then see directly what is happening, and be able to make necessary changes if possible.

In summary, the main goal for the radar systems in the simulation is to gather information that is happening in the lane system and the surrounding area through scanning and reporting any detect behaviors. The structure and performance that were designed was intended to imitate real life radar systems. In addition, these radar systems are dynamic, meaning their locations and specific features can be changed in each simulation based on their designated purpose and the goal of the simulation.

5.2.2 NAB Analysis

The two NAB measures given previously allow the discrimination of nominal from anomalous flight trajectories in almost all cases. This is due to the fact that anomalous flights, generally speaking, do not stay near the lanes nor do they fly in the same direction as the nearest lane. However, trajectories (i.e., x,y,z,t 4-tuple sequences) are of variable length depending on the distance of the flight and the sampling rate. Thus, in order to compare trajectories, it may be necessary to normalize the length of each trajectory to some standard length.

The nominal flights can be distinguished from the anomalous flights by means of a simple feed forward neural network. First, the trajectory lengths are normalized. Next, the NAB measures are computed at every point on the trajectory, and finally, the measures are concatenated into one vector (in this case, distance measure followed by cosine measure). A trajectory generator is created for each flight type based on random launch–land sites (uniformly selected over flight area) and appropriate parameters for the type of flight. Noise is added to the trajectory as follows (the same type of noise is added to all trajectories). First, the ideal trajectory is created. Then starting with the first point and moving to the second point, the error is defined by a circle around the goal point (the circle in the plane normal to the vector from the first point to the second point). A point in the circle (uniformly selected) is chosen as the target point. Next, a point on the line between the starting point and the circle point is chosen using a half Gaussian distribution centered at the circle point; this is the next point in the modified trajectory. When the circle has radius zero, and the Gaussian has zero mean and variance, then the resulting trajectory is the same as the original.

A set of 100 sample trajectories was generated for each flight type, including nominal, for a total of 600 trajectories; half of these were used to train the network to classify nominal versus anomalous flights (two classes), and half were used to validate the result. Figure 5.19 shows the training performance (from Matlab) as well as the 100% correct classification results on the test set.

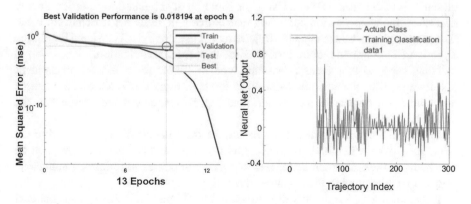

Fig. 5.19 Results of feed forward neural net classification of flight trajectories into two classes. nominal (first fifty) and anomalous (remaining 250). On the left is the network learning performance data and on the right is the classification result

Table 5.1 Classification
results. The 6 trajectory types
are (1) nominal, (2) hobbyist
type I, (3) hobbyist type II,
(4) hobbyist type III, (5)
rogue type I, and (6) rogue
type II

	1	2	3	4	5	6
1	100	0	0	0	0	0
2	0	92	0	3	2	3
3	0	0	100	0	0	0
4	0	4	0	83	10	3
5	0	0	0	0	100	0
6	0	14	0	0	0	86

Once an anomalous flight has been identified, it is possible to develop more refined and model-based techniques to distinguish between the sub-classes. Some characteristics of anomalous flights are:

- *Hobbyist Type I:* not on lanes, not in correct direction, change of altitude in non-vertical direction, launches and lands near the same site
- *Hobbyist Type II:* not on lanes, not in correct direction, change of altitude in non-vertical direction, launches and lands near the same site, hovers for short periods of time
- *Hobbyist Type III:* not on lanes, not in correct direction, change of altitude in non-vertical direction, launches and lands near same site, makes circular motion
- *Rogue Type I:* not on lanes, only goes up, over and down, middle segment may not align with lane, may not be at normal lane altitude, launch and land sites may not be near lanes
- *Rogue Type II:* not on lanes some of the time, not in correct direction some of the time, lanes followed may not be connected in lane network, some changes of altitude not vertical

These characteristics are used to develop models of the various trajectories, and a classifier is built based on them. Using the same set of simulated trajectories already described, the classification confusion matrix given in Table 5.1 is achieved. From these results, it can be seen that the trajectories of the Hobbyist Type I and the Rogue Type II are similar and require further refinement for discrimination.

Now, suppose that there is another type of flight, for example, an amateur just flying circles above some location in the city. Figure 5.20 shows an example trajectory with respect to the ground, and Fig. 5.21 shows it with respect to the airways.

Figure 5.22 shows the distance measure for the amateur flight compared to the noisy data from a nominal scheduled flight. It is clear that the error for such an unscheduled flight is much larger and that flying in circles introduces a noticeable frequency component to the distance signature.

Next, consider the trajectory direction measure. Figure 5.23 shows the signatures for the noisy trajectories of the regular flight. Although it is difficult to see, the mean angles for these are: 26, 31, and 59° for variances of 0.1, 1, and 10, respectively. Now consider the signature for the amateur flight (see Fig. 5.24). The values are all close to 0 indicating that the direction is usually perpendicular to the lane direction.

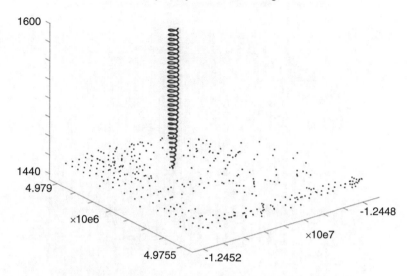

Fig. 5.20 Amateur trajectory flying in circles with respect to the ground

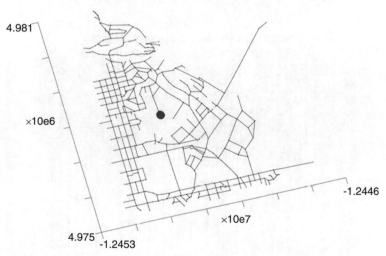

Fig. 5.21 Amateur flight with respect to airway network

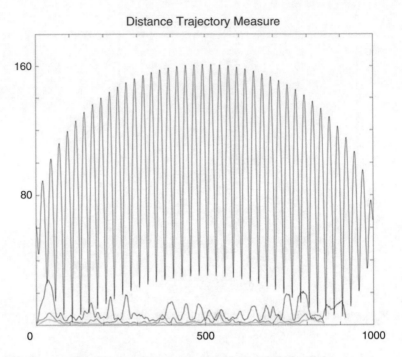

Fig. 5.22 Amateur flight distance signature compared to those of noisy flights

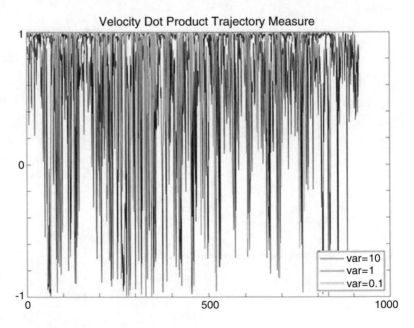

Fig. 5.23 Noisy flights vector cosine w.r.t. lanes

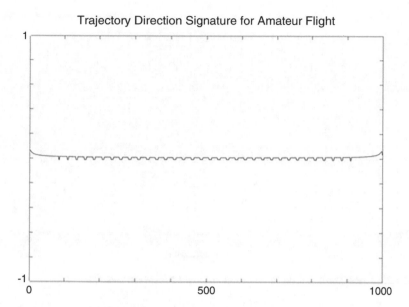

Fig. 5.24 Amateur flight vector cosine w.r.t. lanes

This make sense in that for the first few hundred feet, the lane is directed up, while the circular flight is mainly in the x or y direction.

This particular amateur flight is always near a single lane. Figure 5.25 shows the distance signature from a more densely packed lane area as shown in Fig. 5.26, and a similar oscillating distance is observed.

Next, consider a rogue flight that simply takes the shortest route between launch and land sites (see Fig. 5.27). Figures 5.28 and 5.29 show the distance and direction errors signatures, respectively.

5.3 Next Steps

The lane-based UAS traffic management approach supports efficient and effective trajectory analysis of UAS flights in the airspace. This allows the straightforward detection of unplanned flights through the airspace without having to compare to every existing flight at the time of occurrence. In addition, it is possible to distinguish different types of rogue flights according to the trajectory distance and direction measures.

Several new avenues of investigation are under consideration:

- model Updates due to dynamic lane creation and deletion
- How to exploit knowledge of UTM parameters (e.g., UAS speed limits, lane network topology, 3D corridor constraints, etc.)

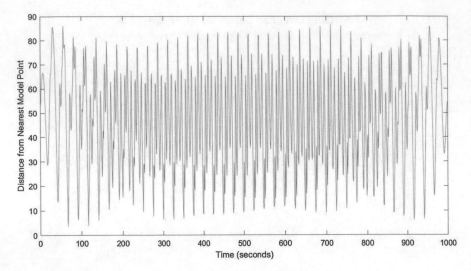

Fig. 5.25 Amateur flight distance signature in a more dense lane area

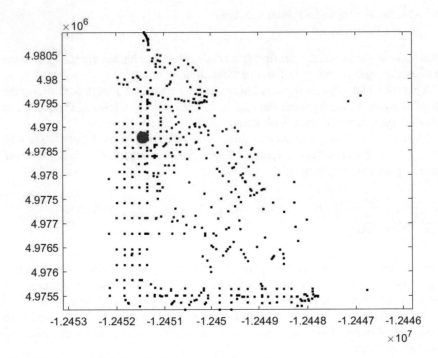

Fig. 5.26 Amateur flight location in a more dense lane area

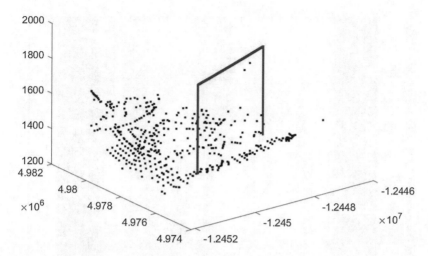

Fig. 5.27 Example rogue flight path

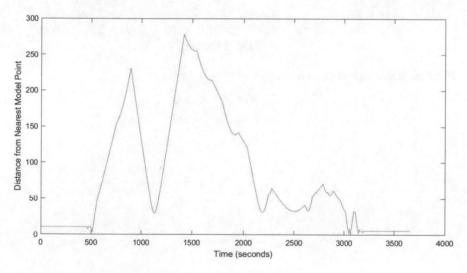

Fig. 5.28 Rogue flight distance error

- Any influence on trajectory measures due to weather, congestion, or other environmental or contingency effects
- The constraints on sensor data to ensure effective identification of anomalous flight patterns
- The role of communications in UAS flight trajectory analysis

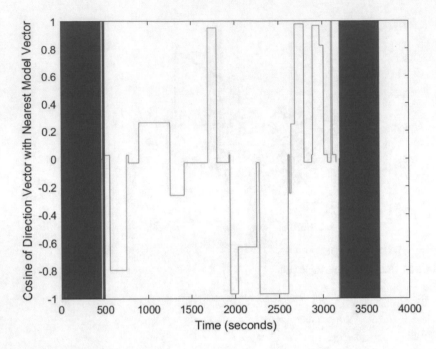

Fig. 5.29 Rogue flight direction error

Chapter 6
UAS Belief–Desire–Intention Agent Architecture

6.1 Introduction

The coming wave of autonomous, semi-autonomous, and human-operated vehicles in low-altitude airspace over densely populated areas requires a new system to automate air traffic control (ATC). The current system relies heavily on the intuition of human pilots and controllers who benefit from over one-hundred years of recorded trials and errors. Even after millions of test cases per year, contingencies occur that confound experts and result in disastrous outcomes (e.g., the failure of coördination that resulted in a mid-air collision over Uberlingen, Germany in 2002 [41]). The computational intractability of enumerating all possible sequences of actions and outcomes that lead to contingencies is the root cause for these disasters; if engineers had known the Uberlingen scenario was possible, they would have avoided it (after the accident, TCAS software was patched to handle it, albeit at great expense due to the complexity of the software and subsequent testing [56]). However, the density and dynamism of the anticipated low-altitude air traffic mandates an automated approach; it is difficult to imagine human controllers managing the separation of thousands of flights per hour. This is the conclusion of most professionals in aerospace across the United States, particularly the Federal Aviation Administration (FAA) and the National Aeronautics and Space Administration (NASA), as well as in Europe and Asia, where UAS Traffic Management systems are being developed rapidly. However, if current ATC systems still experience contingencies after a hundred years, and millions of flights per year, what hope do engineers have in constructing a safe automated traffic management system? This question lies at the heart of this research, and it not only applies to the safe coördinated access of airspace (i.e., maintaining safe separation between aircraft), but also to a plethora of other issues that air traffic controllers face.

The issues that require human intervention, and that make experts nervous about automated air traffic systems, are typically contingency scenarios. Situations

D. Sacharny, T. C. Henderson, *Lane-Based Unmanned Aircraft Systems Traffic Management*, Unmanned System Technologies, https://doi.org/10.1007/978-3-030-98574-5_6

that require safe separation *and* unplanned priority, for example medical rescue, temporary flight restrictions (TFR) due to wildfires, low-fuel, electronics failure, etc. The scenarios, and the combinations or sequences of events that lead to them, are difficult to enumerate, and therefore difficult to plan for. They are *difficult* because in a concrete sense, the determination of whether a solution exists to return to a nominal state may be as difficult as, or likely more difficult than, the Satisfiability (SAT) problem. In the language of computer science, the public relies on pilots and controllers to *heuristically* search for solutions that maintain our safety. It is likely that humans perform these searches at a high level, using abstraction (and perhaps analogies) to reduce the space of possible solutions. So far, no one has shown that the brain, or any part of the nervous system, routinely and exactly solves NP-complete problems [32]. For this reason, the problem of air traffic control, when considering the automation of what human operators are currently responsible for, falls squarely within the purview of computer science. This problem is fundamentally an issue of cognition and computation.

Considering the cognitive and computational nature of the UTM problem, a good strategy for constructing a system to replace human pilots and controllers becomes clear: reduce computational complexity on all fronts. A direct effect of this strategy is the reduction of the number of possible contingencies because by definition there are fewer states to consider (and by implication, fewer undesirable states). It is our contention that this strategy should be executed via two channels. First is through structure and deconfliction; since safe separation is a constraint that must be satisfied in any contingency scenario, it serves to reason that this problem should be easy to solve, and hence low complexity. This is the foundation that the lane-based approach provides. Second, complexity in cognition should be reduced via abstraction, which we explore using agent based modeling and simulation (ABMS), and the Belief–Desire–Intention (BDI) architecture. With this strategy, we expect that the resulting system design will be more robust in the face of contingencies than anything else currently proposed.

6.2 Knowledge Representation and Inference

A UAS agent must analyze a combination of heterogeneous information expressed in logical form (as sentences or statements), computational form (as numerical models of physics or other processes), and sensor data (as measurements from transducers). Each of these forms has its own way to describe uncertainty or error: e.g., frequency models, algorithmic truncation, floating point round-off error, Gaussian distributions, etc. We use our Probabilistic Sentence Satisfiability (PSSAT) method (called *NILS*) that receives information as logical sentences from humans, simulations (e.g., weather or environmental predictions), and sensors (e.g., cameras, weather stations, microphones, etc.), where each piece of information has an associated uncertainty; *NILS* then provides responses to user queries based on PSSAT that determines a coherent overall response to the query and the probability of that response; this new method avoids the exponential complexity of previous approaches. In addition, *NILS* can be used to identify concrete mechanisms (proposed actions) to acquire new data dynamically in order to reduce the uncertainty of the query response. The basis for this is a novel approach to probabilistic argumentation analysis [47].[1]

Most knowledge-based systems do not incorporate a broad notion of uncertainty quantification (UQ), although such a capability would allow a UAS to make more informed decisions, or to acquire more data before coming to conclusions. In addition, it would be better if system responses provided an explanation of how they were derived, as well as how the uncertainty was determined. This can be the result of sensor error, computational error, human error, etc., and the best models should be selected at each time step in order to reduce the variance on quantities of interest. In addition, UAS operating in a lane-based UTM should generate dynamic path planning solutions that can include constraints on time, energy, or uncertainty reduction. The automatic generation of constraints arising from the various models can be used to inform the deployment of data measurement systems. The application studied here is UAV (Unmanned Aerial Vehicle) surveillance and reconnaissance in urban areas. Some work has been done in this general area (e.g., see [58] for a novel guidance law in windy urban environments combining pursuit and line-of-sight laws, and [109] for a multi-cost UAV mission path planner).

We describe here two major novel research results: (1) the combination of formal probabilistic logic methods with state-of-the-art physics-based uncertainty quantification methods and (2) uncertainty driven active information data acquisition, demonstrated by UAV path planning, to optimize performance or to resolve contradictory information. The probabilistic logic method is a re-formulation of the approach described in [71] (although Boole [18] first proposed it); see [46] for details. Basically, Nilsson's method requires first solving the SAT problem (i.e., find all consistent truth assignments) to set up a set of linear equations and then the use of numerical methods to solve them. We, on the other hand, create a set of nonlinear equations and solve them directly [46].

We have previously applied this framework to geospatial intelligence systems in an implementation called *BRECCIA* [95]. *BRECCIA* is designed using a well-documented multi-agent, Belief–Desire–Intention (BDI) framework called Jason [21]. Jason includes an interpreter for an extended version of the AgentSpeak(L) [77] language, which provides a Prolog-like grammar. Both the style, resembling a natural language application, and the operational semantics of the extended language that enable a data-driven architecture fit well with the proactive and reactive goals of *BRECCIA*. Each agent in the *BRECCIA* system is composed of a backward chaining inference module (see [72] for a formal justification of modularity in BDI programming languages) with a probabilistic logic component. This module, and probabilistic component, forms the most abstract fusion implementation in the

---

[1] This material is based upon work supported by the Air Force Office of Scientific Research under award number FA9550-17-1-0077 (DDDAS Geospatial Intelligence).

BRECCIA system. To elaborate, this system may include many functional fusion modules that agents specialize for a particular application. For example, an agent that specializes in managing a particular Unmanned Air Vehicle (UAV) requires a fusion module for estimating target detection. However, the human agent, a "user" in *BRECCIA*, requires information in more abstract terms to support the decision process. For example, a user may query the system by asking for the probability of mission success. Once the list of probabilities has compiled, the nonlinear probabilistic logic algorithm is executed to calculate the probability for the inferred belief. Finally, the inferred belief is added to the agent's knowledge base. The process of belief-revision (when probabilities of antecedents change) currently utilizes the same algorithm, except first a list of justifications, stored in the belief annotations, is compiled until the most abstract inference is located. Belief-revision in the *BRECCIA* system is an active area of research.

6.2.1   Probabilistic Logic

Here we address the problem of finding a suitable representation for uncertainty associated with logical sentences. Although several approaches have been proposed in the past (see [4, 31, 33, 44, 48, 61, 81, 111]), they generally have some significant drawbacks. Usually, these have to do with the computational complexity of the semantics of the sentences (i.e., finding the set of consistent truth assignments is exponential in the number of sentences, or for Domingos, exponential in the number of cliques in the Markov graph [15]).

We have developed a new approach that computes the probabilities of the atoms in the sentences and, in terms of these, provides a solution for $Pr(Q \mid KB)$, where Q is the query and KB is the knowledge base set of sentences (see [46] for details). Moreover, the knowledge of the probabilities of the atoms allows us to determine where the most uncertain part of the argument lies and to allocate resources to lower that uncertainty, thus decreasing the uncertainty of the query. This is done by exploiting the probability of a disjunctive clause and developing a set of equations from the sentences and their probabilities, and then solving those equations (the number of equations equals the number of sentences).

Our approach to probabilistic logic starts with an analysis of Nilsson's method [71].[2] Given a set of n sentences (conjunctions from a Conjunctive Normal Form well-formed formula), $S = \{S_1, S_2, \ldots, S_n\}$, in the propositional calculus, where $\{S_1, \ldots, S_{n-1}\}$ is the KB and S_n is the query, he first finds the set of models of the sentences (i.e., the set of truth value assignments to the sentences that are consistent using the general semantic tree [59] for a set of sentences). In our new approach [46],

---

[2] Note that Nilsson's method for propositional calculus is the same as that proposed by George Boole in the 1800s [18, 19].

we trade the exponential complexity of most other algorithms for the convergence cost of solving for the logical variable (atom) probabilities directly as follows.

First, we assume that the sentences are given in Conjunctive Normal Form. This means that each sentence is a disjunction of literals (a literal is an atom or its negation). Our second assumption is that $Pr(P \wedge Q) = Pr(P)Pr(Q)$ (i.e., they are independent variables); note that if this assumption is violated, our methods also allow the bounds on the probability to be determined. Next, we find the set of logical atoms (i.e., variables) in S; let $A = \{A_1, A_2, \ldots, A_k\}$ be this set. In this case, the probability of a sentence can be computed from the probability of its literals as follows:

$$Pr(L_1 \vee L_2 \vee \ldots \vee L_p) =$$

$$Pr(L_1) + Pr(L_2 \vee \ldots \vee L_p)$$

$$-Pr(L_1)Pr(L_2 \vee \ldots \vee L_p),$$

where the probabilities of clauses on the right hand side are computed recursively.

Assuming that the logical (random) variables are independent, each sentence gives rise to a (usually) nonlinear equation defined by the recursive probability of the disjunctive clause as defined above. The resulting set of equations can be solved using standard nonlinear solvers (e.g., *fsolve* in Matlab) and a set of consistent values for the probabilities of the atoms determined. Of course, one problem with the nonlinear solver approach is that it may not find a solution, even when one or more exist. Thus, our current approach is to solve all equations that have a single unknown (recursively) and then use an iterative method to find a set of atom probabilities that produce the correct sentence probabilities.

6.3 Reinforcement Learning

Reinforcement learning provides engineers with a tool for generating complex decision making programs from high-level requirements. Traditionally, engineers receive high-level requirements, such as "UAS should avoid rain" and then proceed to define all the behavioral rules necessary to fulfill that goal. The effort required to define this program logic is a complex function of the software technology, development cycle, and requirements. Estimating the effort required is itself a large topic of software engineering research [17], and changes to requirements or logic errors can have dramatic costs. Reinforcement learning, however, does not require a manual development of the program logic for the desired behavior. It does require a careful definition of possible states, available actions, and rewards, but it is not necessary to consider all the possible combinations of states and actions.

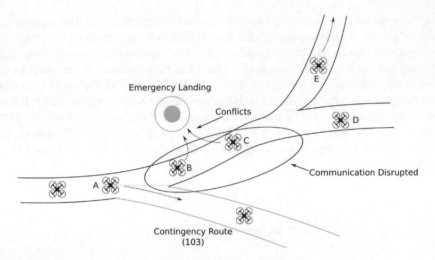

Fig. 6.1 Example scenario: ground communication disrupted for multiple UASs

Consider a scenario in which ground infrastructure supporting UAS communications is disrupted during normal operations (see Fig. 6.1). Currently, the published protocol for handling this contingency is for every UAS to fly back to base if communications cannot be re-established within a given amount of time [10]. Since this is a pre-defined policy, it is worth considering whether such a policy is robust. For example, depending on how many UAS communications are disrupted, the number of conflicts that result from the simultaneous replanning of multiple agents may have negative cascading effects [52]. As the complexity of the UTM system increases, it becomes harder for experts to enumerate all the failure modes and effects; assigning liability and performing post-failure diagnoses will also be difficult.

A comparison between Traffic Alert and Collision Avoidance System (TCAS), which is currently in widespread use by airlines, and a new system called the Airborne Collision Avoidance System X (ACAS X) offers a compelling analogy. TCAS has been described as an "ad hoc rule-based specification" [56]. Limits to its robustness arise primarily because programmers are unable to anticipate the spectrum of operational scenarios, one of which caused a collision over Uberlingen, Germany in 2002. ACAS X, in contrast, adopts a process of modeling and optimization that improves robustness. Kochenderfer describes an early prototype of ACAS X, in which the anti-collision problem is formulated as a partially observable Markov decision process (POMDP) [56]. In this way, collision and alert preferences are treated as inputs and the system logic as an output.

6.3.1   Complexity and Cognitive Structure

When the number of possible states and actions is large, it forces engineers to carefully create abstractions of states and actions (e.g., object-oriented software); otherwise, the program logic becomes fragmented across a large and flat organizational structure. A fragmented program is undesirable because the conceptual links between a high-level requirement and low-level actions are buried in the program logic. On the other hand, a hierarchical structure of states and actions encodes conceptual links explicitly and enables efficient searching through a categorical index (e.g., *Desires* in the BDI architecture).

The Belief–Desire–Intention architecture [40] is a hierarchical organization of states and actions (grouped into *plans*) that was designed specifically for agent models. The architecture not only defines the conceptual structure of a program, but also a process structure that enables dynamic planning, similar to a Markov Decision Process (MDP) [79, 104, 105, 107]. Organizing the program in this way supports both reasoning by the autonomous agent as well as reasoning by human operators. The structure of desires, intentions, beliefs, and plans coincides well with the reasoning of the human operator. For example, a human observer could ask a BDI agent directly, "What is your current plan?", and the agent could respond, "Correcting my heading to get back in the lane."

6.3.2   Complexity and Airspace Structure

In previous work, we outlined the structure and analytical capabilities made available by the lane-based airspace structure [89, 91]. From a cognitive perspective, the environment in which autonomous agents operate is dynamic and uncertain, and agents are resource-constrained and have only a local view of the world. The lane-based structure provides autonomous agents with more information about the state of the airspace while requiring less computational effort to reason about it. The primary basis for this is that agents only need to consider the schedule on each lane, as opposed to the entire trajectory of other aircraft in a free-flight airspace structure.

In the free-flight model, where any trajectory is allowed, agents must sample other trajectories and their own to ensure safe separation. The sampling resolution necessary to ensure safe separation, i.e., the discretization of trajectories, depends on how trajectories are specified and the available time and resources to perform the sampling. Each agent must perform this computation every time it considers a new or altered path. The lane-based approach, however, pre-calculates safe separation in the spatial domain, so agents only need to consider the schedule.

6.3.3   Learning and BDI

UAS agents have a Belief–Desire–Intention architecture that functions as shown in Fig. 6.2a. The BDI cycle involves updating the beliefs, determining the desires, choosing an intention (goal), and then selecting an appropriate plan to achieve that goal. A *belief* is represented as a disjunction of logical variable literals, and the entire belief set is a conjunction of such beliefs (in Conjunctive Normal Form (CNF)). A *desire* is a belief that the agent would like to make true, and an *intention* is a belief that is a current goal (of which only one is active at a time). The selection of a plan is the action at the cognitive level, and optimal cognitive policies pick the best plan for a given state. Focus 1 in Fig. 6.2b is where the cognitive learning takes place; i.e., the actions are a set of possible plans to achieve a specific goal, and an optimal policy chooses the best plan for a given state. The selected plan is then executed until either completion or preemption. Focus 2 concerns policies at the physical level (i.e., moving through space).

For example, given cognitive states $S = \{S_1 \equiv GoToDestination - NoDrift, S_2 \equiv GoToDestination - Drift, S_3 \equiv Fail, S_4 \equiv Succeed\}$, where there may be several plans to achieve a goal (e.g., shorter or safer routes), and the action is to select one of these plans. Rewards are associated with states and actions,

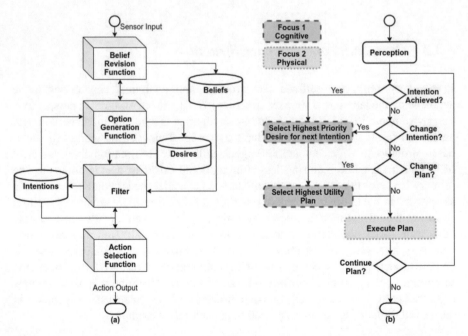

Fig. 6.2 (**a**) BDI architecture (taken from [50]). (**b**) Reinforcement learning Focus 1: cognitive-level plan selection to achieve goal, and Focus 2: actions in the individual plan (this figure is adapted from [23])

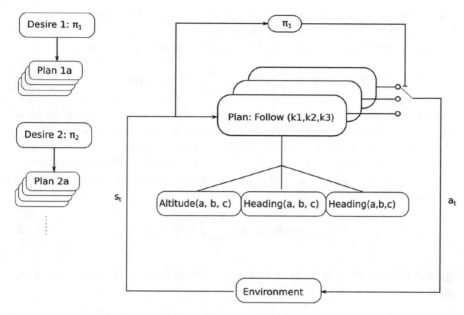

Fig. 6.3 MDP representation of BDI architecture (this figure was adapted from [9])

and reinforcement learning is applied to find optimal policies. This is done over a large number of environmental and UTM conditions. Actions are parameterized by considerations like estimated required time, risk, communications connectivity, etc.

At the physical level, a plan may consist of a sequence of lanes with associated entry–exit times. Alternatively, a plan may consist of a sequence of GPS waypoints and times. We have already performed a preliminary study of this aspect of UAS plan optimization (see [88]), and shown how optimal policies (for moving through space) can be determined in the context of environmental conditions (e.g., wind). Cognitive-level reinforcement learning follows the same process as traditional reinforcement learning (Fig. 6.3) [101].

6.3.4 Experiments

6.3.4.1 Learning at a Low Level

The goal here is to determine an optimal action selection policy for a UAV with a given destination goal and a set of specific environmental conditions that defines the state space. The agent must operate successfully in this environment by learning a utility function on the state of the world and from those utilities determine an optimal action policy for each state. We consider an agent in a fully observable environment. Once a policy, π, is learned to maximize utility, $U(s)$, then it deterministically specifies the action for each state, and the agent will always choose action $\pi(s)$ for state s. The goal is to maximize the expected utility (see [86] for details). The utility for each state is defined by the Bellman equation:

$$U(s) = R(s) + \gamma max_{a \in A(s)} \sum P(s' \mid s, a)U(s') \qquad (6.1)$$

where $U(s)$ is the utility of state s, a is an action, $A(s)$ is the set of actions possible for state s, and P is the probability of state s' given state s and action a. We use value iteration to solve for the state utilities; i.e., the above equation is iterated, updating the utility of each state until convergence is achieved. Once the utilities are known, the optimal policy at each state corresponds to the action that maximizes the expected utility from the action:

$$\pi^*(s) = argmax_{a \in A(s)} \sum_{s'} P(s' \mid s, a)U(s') \qquad (6.2)$$

State Representation   In order to solve for the optimal policy for UAV control, we define the state space as

$$S = \mathcal{Z}^3 \times \Re^3 \times \Re^+ \times \Re$$

where the space is composed of three integer grid coördinates, a real-valued 3D wind vector (whose magnitude is the wind speed), a precipitation value, and a temperature value. Note that although the wind, precipitation, and temperature values have different dimensions, their values are represented by indexes that designate intervals in the appropriate range. For the study here, the grid consists of a 4×4×4 set of voxels (representing airspace volumes where the specific dimensions of the air volumes are determined by the problem under consideration; here we assume reasonably large volumes), 2 values are used to indicate the wind (none, high), precipitation is binary (raining or not), and 3 values for temperature (cold, normal, hot). Thus, the state vector is a 6-tuple, where the number of values for each element is [4,4,4,2,2,3], resulting in a total of 768 distinct states.

Possible Actions   The action set for this problem is simply the selection of a neighboring air volume; in particular, one of the 6 orthogonal direction cells. These actions will be labeled $\{X, -X, Y, -Y, Z, -Z\}$, so that these actions align with a standard frame in the center of the cell (see Fig. 6.4).

State Transition Function   Next, a probabilistic state transition function, $P(s'|s, a)$, is defined, which provides the probability that state s' results from choosing action a while in state s. The particular function used here is based on the physics of the state transitions, accounting for the impact of motion based on wind, temperature, and precipitation.

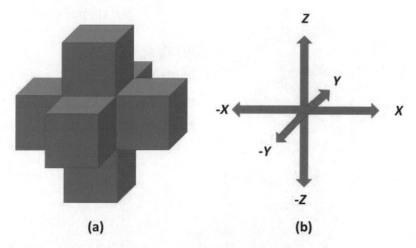

Fig. 6.4 Actions for UAS low-level reinforcement learning. (**a**) Airspace volumes. (**b**) Action directions

Reward Function Finally, a reward function is defined as follows:

$$R(s) = \begin{cases} -0.04 & s \neq \text{goal,excluded state} \\ -1 & \text{excluded state} \\ +1 & \text{goal state} \end{cases}$$

The goal state has grid location [4,4,4] and reward value 1, while excluded cells have a value of -1.

Results In order to better understand the method, a specific example will be considered here. First, a $4 \times 4 \times 4$ grid of 64 air volumes as shown in Fig. 6.3 will be indexed by either their grid coördinates or by a simple index. E.g., cell [3,2,4] will also be identified by the index 31. In addition to the grid location, each volume also has temperature, wind, and precipitation information. This latter aspect will be considered below.

The actions available are directly tied to moving to one of the neighboring (closest) six neighbors. Figure 6.5 shows the states. Note that when a cell is on the boundary, then the UAV is not allowed to exit the $4 \times 4 \times 4$ grid, and so if that direction is chosen, the UAV will remain in the same cell with some probability.

In order to solve the value iteration problem, the neighbors of each cell in each action direction are first determined. A few of these are given in Table 6.1.

Next it is necessary to define the probability of moving into each neighboring cell given the desired action. This is provided in terms of Table 6.2. Note that we assume it is more likely that motions in the X–Y plane will have a certain probability and that moving up is more uncertain than moving down. Although we have assigned likely values here, these are also parameters that may be learned over time.

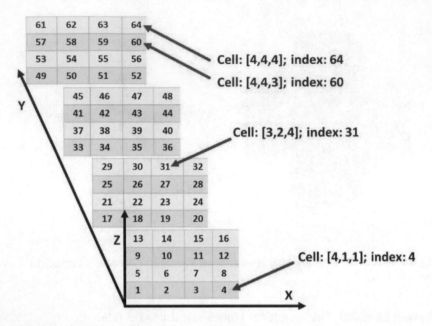

Fig. 6.5 States for UAS low-level reinforcement learning

Given this information, it is possible to run the value iteration algorithm and find the utilities for the states. Figure 6.6 shows the utilities produced at each state as well as the path through the highest probability sequence of states. Note that this may not match the optimal actions selected since it does not take into account the maximal expected utility of the action. It does show however that the UAS will most likely move up and then over in such a way as to avoid the excluded air volume (index 60, cell [4,4,3]). Figure 6.7 shows how the utility values converge for this problem.

An optimal policy can then be determined for each state and produces the result shown in Table 6.3. These results are also shown in Fig. 6.8. Note that the red arrows indicate a move in the Y direction. Of course, the actions are not deterministic, and in order to better understand the impact of this policy, 1000 trials were run with start location cell [1,1,1], index 1, and with goal location cell [4,4,4], index 64. One cell is excluded, cell [4,4,3], index 60; if the UAS enters that cell, it must land and terminates the mission. Figure 6.9 shows the number of times each cell in the airspace was traversed by the 1000 trials. As can be seen, much information may be gleaned from these results as to the probability that a UAS will be in the assigned air volume (we have not included temporal aspects, but that is readily available, if desired). Also, note that a half a percent of the trials resulted in failure.

Next, consider the impact of a stiff wind blowing in the Y-axis direction. This information is easily added to the model by simply providing the state transition probability for motions impacted by the wind. This is called context-based probabilistic state transition. In the case of a strong wind in the Y-axis

Table 6.1 State transitions for UAS low-level reinforcement learning

State Index	X	-X	Y	-Y	Z	-Z
1	2	1	17	1	5	1
2	3	1	18	2	6	2
3	4	2	19	3	7	3
4	4	3	20	4	8	4
5	6	5	21	5	9	1
6	7	5	22	6	10	2
7	8	6	23	7	11	3
8	8	7	24	8	12	4
9	10	9	25	9	13	5
10	11	9	26	10	14	6
...						
61	62	61	61	45	61	57
62	63	61	62	46	62	58
63	64	62	63	47	63	59
64	64	63	64	48	64	60

Table 6.2 State transition probabilities for UAS low-level reinforcement learning

Action	X	-X	Y	-Y	Z	-Z
1	0.60	0.00	0.10	0.10	0.05	0.15
2	0.00	0.60	0.10	0.10	0.05	0.15
3	0.10	0.10	0.60	0.00	0.05	0.15
4	0.10	0.10	0.00	0.60	0.05	0.15
5	0.15	0.15	0.15	0.15	0.40	0.00
6	0.05	0.05	0.05	0.05	0.40	0.80

direction, say produced by afternoon canyon winds in Salt Lake City, a set of state–action probabilities are provided (either by learning over time or by physics-based simulation). Figure 6.10 shows the values used here.

Running value iteration with these probabilities gives rise to the convergence values shown in Fig. 6.11. The policy determined for these utilities is shown in Fig. 6.12. Some interesting observations may be made. For example, the policy never chooses a Y-axis action. Figure 6.13 shows the number of times each airspace volume is traversed over 1000 trials. Note that there are a significantly higher number of failures (39) due to the strong wind. Of course, this policy is the result of

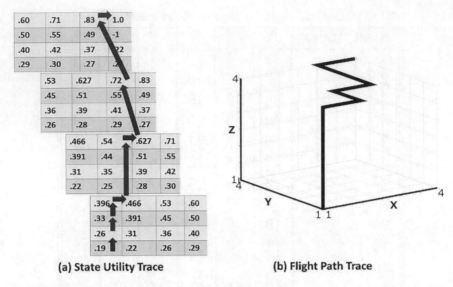

(a) State Utility Trace   **(b) Flight Path Trace**

Fig. 6.6 The state utilities and path through the highest utility states. (**a**) State utility trace. (**b**) Flight path trace

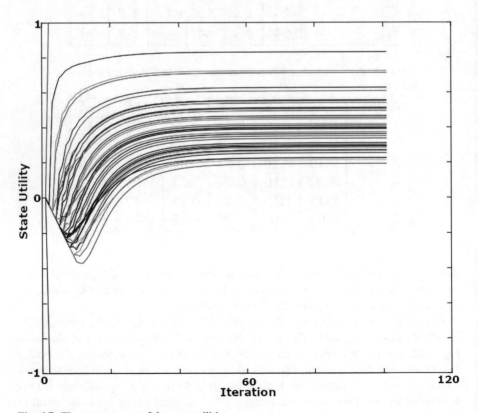

Fig. 6.7 The convergence of the state utilities

Table 6.3 Optimal policies for the states

State	Policy	State	Policy	State	Policy
1	5	23	5	45	1
2	5	24	5	46	1
3	5	25	5	47	3
4	5	26	5	48	3
5	5	27	5	49	5
6	5	28	5	50	5
7	5	29	3	51	5
8	5	30	3	52	2
9	5	31	3	53	5
10	5	32	3	54	5
11	5	33	5	55	5
12	5	34	5	56	2
13	3	35	5	57	5
14	1	36	5	58	5
15	3	37	5	59	2
16	3	38	5	60	-
17	5	39	5	61	1
18	5	40	5	62	1
19	5	41	5	63	1
20	5	42	5	64	-
21	5	43	5		
22	5	44	4		

a certain level of tolerance for failure vs. energy expenditure. It is possible to vary these parameters and produce policies less likely to fail by entering excluded zones. Finally, it is important to note that the method can be considered having a learning part and an application part.

6.3.4.2 Learning at a Cognitive Level

To demonstrate the BDI reinforcement learning at the cognitive level, a simulation was run to determine the optimal policies for UAS agents in an environment that contains rain and wind. The cognitive state of each UAS before the simulation begins is represented by a knowledge base containing the clauses in Table 6.5. The initial goal is for the UAS to be assigned a mission, so the intention stack is started with the desire to be *ASSIGNED*. Once assigned, the UAS selects intentions from

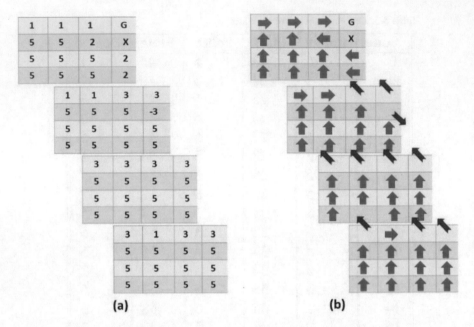

Fig. 6.8 Optimal policies for the states (pictorially). (**a**) Action numbers. (**b**) Directions

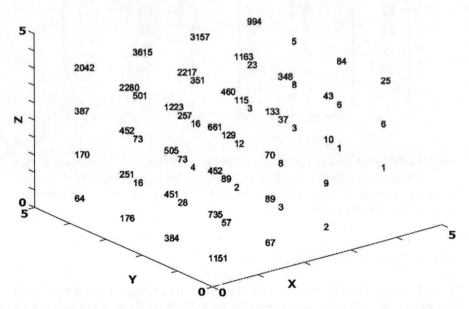

Fig. 6.9 The number of times each cell is traversed in 1000 trials

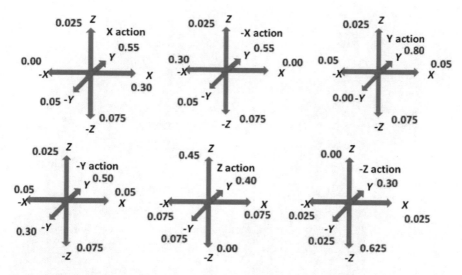

Fig. 6.10 The context-based probabilistic state transition probabilities for the case of a strong wind in the Y-axis direction

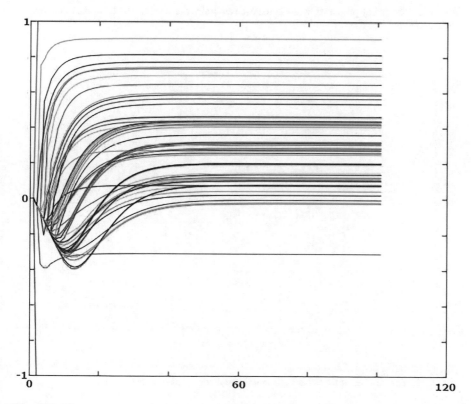

Fig. 6.11 The convergence of value iteration with the context-based method

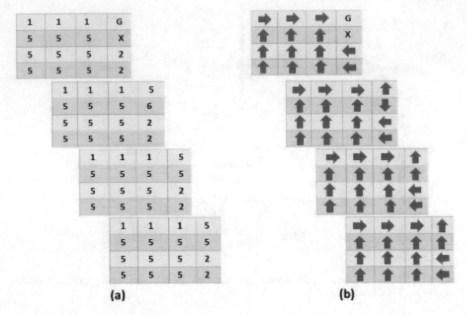

Fig. 6.12 The policy produced by the context-based state transition method. (**a**) Action numbers. (**b**) Directions

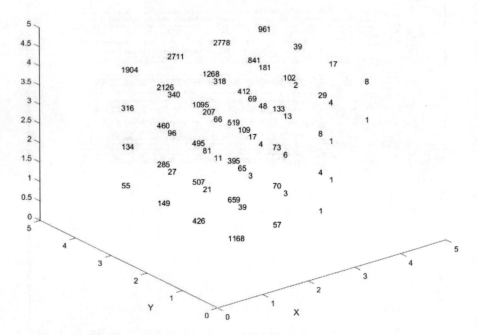

Fig. 6.13 The number of times each space volume is traversed over 1000 trials

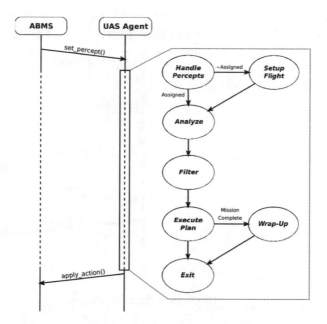

Fig. 6.14 Combined sequence and state diagram showing agent architecture

the precedence list shown in Table 6.6 (i.e., after being assigned, the next desire is to remain *IN_LANE*).

A combined state and sequence diagram outlining a single reasoning cycle is shown in Fig. 6.14. When the UAS agent receives a percept from the simulator, containing both state information as well as messages from other agents, it parses the data and updates its knowledge base. In the *Analyze* state, the agent considers its desires and updates its current intention. The *Filter* state involves selecting the optimal plan for the current intention, and in *Execute Plan*, the required low-level actions are taken.

The process for generating and assigning flights follows the general design proposed by NASA where a UAS Service Supplier (USS) is responsible for deconflicting flights with other USSs in the system. To accommodate the ABMS setup, the USS is also responsible for generating random flight requests and auctioning them to UAS agents. This process is diagrammed in Fig. 6.15.

6.3.4.3   Environment Model

The environment used to train a policy is shown in Fig. 6.16, where solid circles mark areas of rain and dashed circles mark areas of wind. Rain is modeled as a scalar intensity value that decreases with distance from the center of the feature. Rain affects the speed of UAS proportional to the amount of rain. Wind is similarly defined, except it affects both speed and heading of a UAS in a direction tangential to the radius of the wind feature.

Fig. 6.15 Flight path
generation and assignment

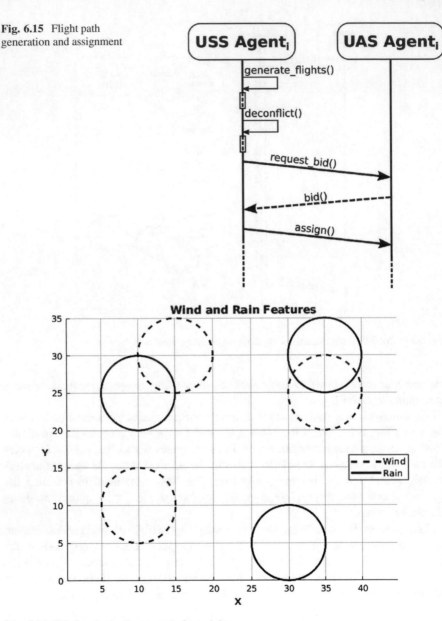

Fig. 6.16 Wind and rain placement in for training

6.3.4.4   Actions

UAS agents in this simulated framework have the ability to set their velocity after
each reasoning cycle. In a scenario without wind or rain, the desires are selected in
order of precedence (lower precedence happens first). However, due to the dynamics

Table 6.4 UAS high-level plans

Plan	Description
CORRECT_HEADING	Heading Optimized Controller
CORRECT_SPEED	Speed Optimized Controller
FOLLOW_LANE	Main Lane Following Control
GO_TO_LANE	Take Immediate Action and Fly to Lane Segment

in the environment brought in by wind and rain, agents are affixed with logic that requires replanning when the situation is not *NOMINAL*. In this case, the agent must choose a contingency plan that returns the cognitive state to *NOMINAL*. There are a number of plans that may achieve this, given in Table 6.4. The selection of plans is determined by a policy obtained through reinforcement learning.

6.3.4.5 Transition Probabilities and Rewards

To generate the transition probabilities, a Monte Carlo simulation was run, and a three-dimensional state–action transition matrix, representing the probability $P(s'|s, a)$, was generated from the data.

The reward model, $R(s, a)$, considers only the eight states generated by the Cartesian product of *NOMINAL*, *RAIN*, and *WIND*, and the four high-level plans in Table 6.4. A state reward (R_s) of +6 was assigned to any state that was *NOMINAL*, and −2 for any state that was not. Plan rewards were set as follows (reflecting their cost to execute):

$$R_a(\text{FOLLOW\_LANE}) = -1$$

$$R_a(\text{CORRECT\_SPEED}) = -3$$

$$R_a(\text{CORRECT\_HEADING}) = -5$$

$$R_a(\text{GO\_TO\_LANE}) = -8$$

6.3.5 Policy Selection

A policy was selected by running value iteration and generating state utilities. A trace of the state utilities after each iteration (Eq. 6.1) is shown in Fig. 6.17. A slice of the transition probability matrix for the plan *FOLLOW_LANE* is depicted by the digraph in Fig. 6.18. The UAS agent then combines the state utilities and the transition probabilities using Eq. 6.2 to select the optimal plan.

A graphical depiction of the behavior of a single UAS mission without rain or wind contingencies is shown in the plan and state graph in Fig. 6.19. Figure 6.20

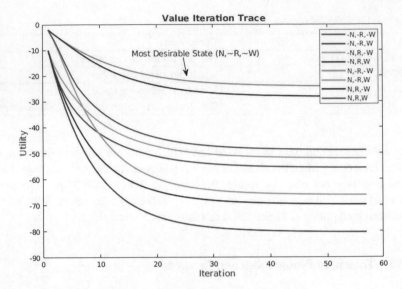

Fig. 6.17   Value iteration trace. N-Nominal, W-Wind, R-Rain

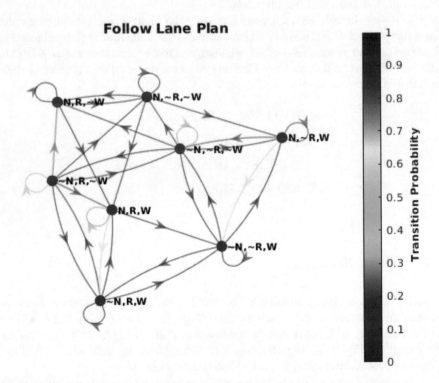

Fig. 6.18   Transition probabilities for *Follow Lane* plan

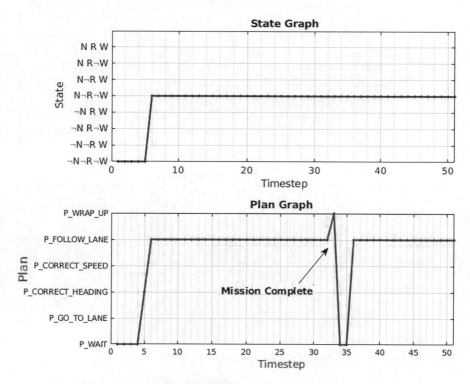

Fig. 6.19 Nominal behavior without contingencies

shows the behavior of a UAS navigating the same trajectory through a rain feature
with a learned policy. Finally, in Fig. 6.21, a behavior trace of a UAS navigating the
rain feature using a programmed policy that deals with the rain contingency directly
by correcting its speed.

6.3.6 Discussion

The experiments demonstrate that reinforcement learning at the cognitive level
is a viable option for programming agents in a UTM system. The program in
this instance was comprised of a number of plans that could be engineered
independently, in contrast to the currently proposed strategy of enumerating risk
factors and developing contingency plans in concert across the industry.

In this simple experiment, the resulting policy is guaranteed optimal with respect
to the rewards because dynamic programming was used. In a large-scale system,
the number of possible states and actions may be too large to pre-calculate utilities.
However, a plan can be engineered in which the UAS agent performs dynamic
programming over a narrowed set of states provided by the *Analyze* step in the

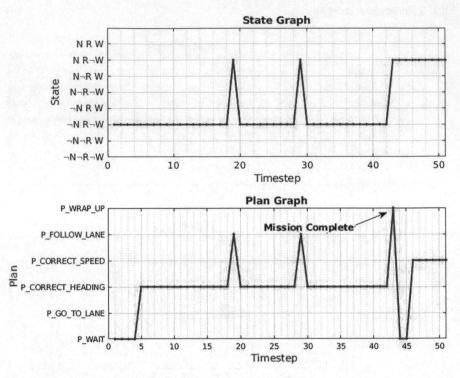

Fig. 6.20 Learned policy with rain contingency

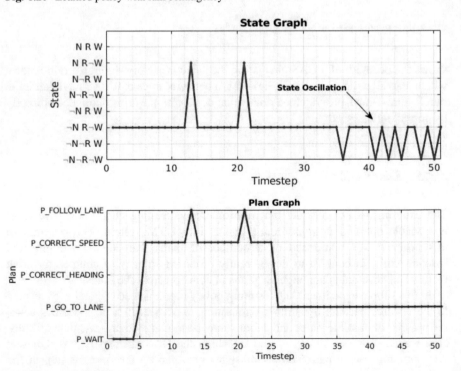

Fig. 6.21 Programmed policy with rain contingency

Table 6.5 UAS KB at initialization

ID	Clause
1	IN_LANE ∧ ON_HEADING ∧ SPEED_OK → NOMINAL
2	LAST_LANE ∧ AT_NEXT_WAYPT → AT_FINISH
3	¬IN_LANE
4	¬ON_HEADING
5	¬SPEED_OK
6	¬ASSIGNED
7	¬IN_FLIGHT
8	¬AT_START
9	¬AT_NEXT_WAYPT
10	¬ADVANCE_LANE
11	¬WRAP_UP

Table 6.6 UAS desires and precedence

Desire	Precedence
ASSIGNED	10
IN_LANE	20
ON_HEADING	30
SPEED_OK	40
AT_NEXT_WAYPT	50
ADVANCE_LANE	60
WRAP_UP	70

BDI architecture. Since the BDI architecture supports replanning when a plan fails or when desires change, the agent can avoid over-committing to a plan that did not consider a particular contingency.

This fact is demonstrated in the experimental output of Figs. 6.19, 6.20, and 6.21. The explicitly programmed policy is brittle in the face of a rain contingency because the speed correction causes the UAS to overshoot lane waypoints. The learned policy selected a different plan, one that corrects heading and speed concurrently and proves to be more robust.

6.3.7 Conclusion and Future Research

We have shown how states and actions at the cognitive level can be combined with reinforcement learning to generate optimal policies for UAS agents. Additionally, the BDI architecture provides a convenient structure for defining high-level states and actions, while reducing the engineering complexity by allowing plans to be designed independently. The benefit of this approach is that it reduces the amount of programming logic required to build robust policies and utilizes a logic structure that supports operational insight since decisions are expressed in a human-understandable form.

In upcoming research, our intention is to demonstrate the performance of the combined cognitive-level reinforcement learning BDI architecture with respect to contingency scenarios where a large number of states exist. In these scenarios, the agent must replan when optimal plans do not produce the correct outcome due to unforeseen states.

Also, the effects of replanning and contingency handling on the aggregate state of all UAS agents in a UTM system must be understood. If it can be shown that the best available policy for UAS agents includes dynamic replanning and the proposed cognitive structure, then this strategy will enable a more rapid adoption of autonomous agents due to the decreased engineering complexity.

Chapter 7
Contingency Handling

7.1 Introduction

NASA engineers have published a number of system requirements in an effort to enable dense operations of unmanned aircraft systems (UAS) in urban environments. These requirements describe a free-flight model, where operators are afforded the maximum flexibility to design individually optimal trajectories, with the caveat that all operations must be strategically deconflicted prior to flight. Strategic deconfliction reduces the probability of having to perform tactical deconfliction using onboard sensors and real-time algorithms. Such approaches require a common protocol to guarantee that UAS do not collide, but do not scale well. Thus, UAS Service Suppliers (USS) must deconflict their planned trajectories pairwise prior to flight in order to achieve strategic deconfliction. However, sometimes flights are not able to follow their plan and must handle some contingency; a contingency is a future possible event, usually causing problems or making further plans necessary. This chapter describes a communication-based protocol to coordinate airspace during flight. This protocol was developed as part of the Air Force Office of Scientific Research program on Dynamic Data-Driven Applications Systems (DDDAS) [58, 67]. In a seminal article describing the purpose and scope of dynamic data-driven applications systems, Darema [30] describes a motivating example where injecting experimental data into a long-running computation (informing oil exploration decisions) could be performed in an online manner to produce better results. An *online* program in the DDDAS paradigm accepts data whenever it is available and could also inform the measurement process to improve system efficiency. The computational effort required to produce good decisions is also a motivating factor for the development of a DDDAS approach to traffic management described here.

NASA and the FAA are making a concerted effort to develop an Unmanned Aircraft System (UAS) Traffic Management (UTM) system to enable large-scale

© The Author(s), under exclusive license to Springer Nature Switzerland AG 2022
D. Sacharny, T. C. Henderson, *Lane-Based Unmanned Aircraft Systems Traffic Management*, Unmanned System Technologies,
https://doi.org/10.1007/978-3-030-98574-5_7

UAS exploitation in urban environments. The UTM is organized in terms of UAS operators who manage their flights through UAS Service Suppliers (USS). These service suppliers must declare the geographic region of their flights (in terms of 4D trajectories of space–time), and moreover, must strategically deconflict their flights pairwise with all other UAS flights in the region (we call this method *FAA-NASA Strategic Deconfliction* or FNSD). This can easily lead to quite complex path planning and coördination problems, and also requires USS to share data which would best be kept private. We have introduced a lane-based organizational structure for a UTM in which a set of lanes are defined (much like a ground road network), and then a USS simply reserves a sequence of lanes from takeoff site to destination site [90, 91]. In that work, we demonstrated a lane reservation system that efficiently guarantees strategic deconfliction, however that only applies to flights that are yet to be active in the airspace. Active flights experience a more dynamic situation, where contingencies can occur.

Contingencies are communicated to agents in an online fashion, either by tactical avoidance sensors such as radar and sonar, or as information from authorities and other agents. Both sources can result in undesirable system responses, for example cascading effects due to high-density operations [52] and unstable control response due to the structure of the information flow [36]. We described earlier the Lane Strategic Deconfliction algorithm (called LBSD) and showed that it has very low complexity, and allows for quite acceptable lane stream properties. Overall, contingencies that lead to a violation of safe separation represent the most critical element to consider in the design of a large-scale traffic management system. Safe separation requires agents to plan collision-free paths, which in the most general case of multiple-agent planning is PSPACE-hard. Even the more narrow problem of tuning velocity profiles is NP-hard [3].

We have given a lane-based airspace model that enables the propagation of contingency information in a well-defined manner. UAS plan locally in real-time within lanes, broadcasting contingencies (as deceleration events) to neighboring lanes that are likely to be effected. Unlike car-following models [70], information from a contingency can reach multiple agents at the same time, yet enabling agents to react in a similarly predictable way. The theoretical contribution here provides an efficient real-time algorithm for strategic deconfliction and applies a solution in terms of ground-delay (delaying access to the airspace network) or air-delay. The experimental section below demonstrates the ability to resolve conflicts within a simulated environment.

7.1.1 Lane-Based UTM

A central issue concerning the DDDAS paradigm is the choice of model, and how information is represented, distributed, and consumed. The lane-based airspace structure is a model for the configuration of UAS in space and time and contrasts with other proposed models, such as the grid-based structure proposed by NASA.

For example, in a grid model UAS share position information (through a USS as a proxy) within cells of a grid, and it is incumbent on USS to determine whether changes to trajectories could impact operations in neighboring cells. In other words, the flow of information between cells is not explicit in the model and represents a major point of uncertainty in the system. This contrasts to the lane-based approach, where impacts of trajectory changes (the dynamic data in this system) within a lane propagate in a well-defined manner throughout the lane network. The lane-based approach imposes a clear downstream and upstream direction to the information flow because lanes form a graph structure that mirrors the possible paths by UAS. The representation of trajectories in the lane-based approach is simple, as described below, and limits the amount of information that must be shared between aircraft to ensure safe separation. Finally, utilities can be defined in a straightforward way for both the UTM and UAS; e.g., the distance between all flights is important for the UTM, while maintaining desired speed and distance to destination characterize the utility of a configuration for a UAS.

Given a set of ground launch and land sites, a set of one-way lanes is defined, which provides a path from any launch to any land site. A lane is a directed 3D vector with its tail as the entry point to the lane and its head as the exit point. A flight path is a sequence of lanes starting with a vertical launch lane and ending with a vertical land lane. A crucial constraint on lanes is that every vertex (entry or exit point) has either in-degree 1 or out-degree 1; this allows the deconfliction of flights by considering lanes as opposed to nodes in the network.

In order for two UAS to be safe, they must at no time be closer than some minimal Euclidean distance, called d_S. We assume that lanes are defined so that no two lanes have points closer than d_S unless the two lanes share an endpoint. Figure 7.1 shows the simple lane layout used in the set of experiments described below. There are 51 lanes, along with 10 launch lanes and 10 land lanes.

7.1.2   Contingencies

Both approaches (FNSD and LBSD) are subject to the problem of contingencies when a UAS flight departs from its nominal plan (e.g., slows down, goes off-course, etc.). Due to the complexity of the UTM system, predicting the effects of contingencies is a major impediment to the wide-spread integration of UAS into the urban airspace. The currently published protocol for mitigating many contingencies requires the UAS to try to return directly to its launch site [10]. However, this trajectory may not be strategically deconflicted and requires obstacle detection and avoidance along the way.

The lane-based model, together with the coördination protocol proposed here, offers a method to mitigate such a contingency and also provides techniques to analyze the possible outcomes of different contingencies. The well-defined structure of lanes suggests that only a restricted set of contingency trajectories need to be considered, those that follow the lane structure and those that do not. For example,

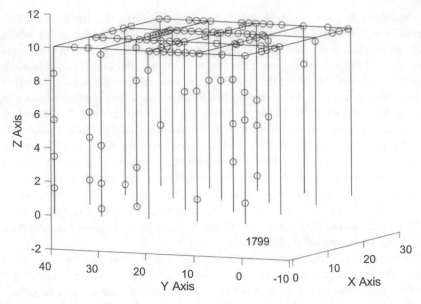

Fig. 7.1 Set of UAS on airways during discrete event simulation. Red dots represent UAS in flight; blue lanes are launch lanes

addressing contingencies where UAS must exit a lane could include designating emergency side lanes where a UAS can wait, or dynamic landing lane creation to go to the nearest safe landing site. In the case that the UAS can still follow lanes, the simulations demonstrated in the experimental section below offer a method to understand the possible outcomes. In [90] an analysis of the impact of lane density on the delay of a requested lane reservation was shown to be an instance of a process of random space filling, sometimes referred to as Renyi's parking problem [80]. The lane-based structure imposes constraints on the network that make this analysis possible and could inform what a safe operating density for the UTM should be.

The proposed real-time tactical deconfliction method described in this paper simply modifies UAS speeds throughout the network in such a way as to avoid conflict. This method effectively absorbs contingencies when the UAS agent is still capable of following lanes. In the event of a contingency where a UAS cannot still follow lanes, the impact is minimized because non-contingent operations remain within the lane structure.

As long as all flights are strategically deconflicted and stick to their assigned flight paths, then there will be no problems in the airspace. However, situations can arise that cause a flight to deviate from its nominal path. Example include:

- **Weather conditions:** wind, rain, snow, etc. can cause a flight to slow down in the direction of the lane or move off-course.
- **Platform failures:** Power, navigation, fuel, or structural damage, etc., can prevent a flight from maintaining its speed, altitude, etc.,

- **Priority preemption:** Emergency or other aircraft may be given authorization to use a lane during the time allocated for a lower priority regularly scheduled flight.

Note that these and other conditions may be statistically predictable, but it is not possible to know exactly when they will occur.

There are a couple of ways to handle contingencies: pre-planned versus dynamic mechanisms. Pre-planned mechanisms involve either modifying the design of the airway lane structure so that flights may address contingencies by using those lanes, or by establishing UTM parameters or policies that allow specific contingencies to be overcome. Dynamic mechanisms are those that create structures on the fly or provide for UAS to interact and modify their flight paths in order to stay safely deconflicted.

7.1.3 Pre-planned Contingency Mechanisms

Consider how ground road networks provide ways for automobiles to handle contingencies. On most roads if a vehicle breaks down or runs out of gas, etc., there is a shoulder or emergency lane where the car can pull over and take care of the problem. Moreover, in town there are usually many less-traveled side streets where a car can pull off the main road. Similar to ground road networks, it is possible to create emergency air lanes alongside every regular lane so that a UAS that has a problem can pull into that emergency lane and address its problem. Such emergency lanes can be defined when the airway lane network is constructed.

If a UAS is unable to continue its flight to destination, then it would need to land. This can also be addressed when the air lane network is defined. For example, every emergency lane can have an associated landing lane. In this case, the emergency side lane could be comprised of two sub-lanes with opposite directions of travel toward each other. A landing lane can have its entry point where the two emergency sub-lanes meet. All this structure can be pre-defined.

Of course, multiple aircraft in the same lane may have problems, and therefore their use of the emergency and landing lanes must be coördinated. If they retain adequate platform control, then the reservation system can be used to schedule them from their current positions through the emergency lane and down the landing lane. Such coördination can take into account the remaining individual capabilities of the aircraft involved. If some UAS is seriously incapacitated, but can still fly (i.e., it is not deploying a parachute and falling from the sky!), then it can be cleared to descend as needed. A multi-craft emergency in the same lane can also be resolved dynamically (see below) by having the flights coördinate between themselves the usage of the emergency and landing lanes.

It is also possible to implement UTM policies to help manage certain types of contingencies as opposed to defining emergency lane structures. For example, if a UTM requires all flights to fly at the same speed, say s, then if some flight slows down to speed s', then the UTM can require all flights to slow to speed s'. At this reduced speed the flights will remain deconflicted. To see this, consider first if two flights are in the same lane when the speed reduction occurs, then the distance between them will not change. Now consider two flights that are in lanes that merge to a single lane. Let flight f_1 be at distance d_1 from the merge point (i.e., the exit point for the current lane for f_1 and the entry point for its next lane), with speed s and scheduled arrival time t_1. Similarly, let flight f_2 be at distance d_2 in its lane from the merge point and moving with speed s and scheduled to arrive at the merge point at time t_2. Suppose that at time t both flights must reduce their speeds to $s' < s$. It must be shown that the headway distance is maintained at the lower speed. For the scheduled flights we have

$$d_1 = s(t_1 - t)$$

$$d_2 = s(t_2 - t)$$

with the reduced speeds we have

$$d_1 = s'(t_1' - t)$$

$$d_2 = s'(t_2' - t)$$

which yields

$$t_1' = t + \frac{d_1}{s'}$$

$$t_2' = t + \frac{d_2}{s'}$$

Substituting the definitions of d_1 and d_2:

$$t_1' = t + \frac{s(t_1 - t)}{s'}$$

$$t_2' = t + \frac{s(t_2 - t)}{s'}$$

Subtracting the first from the second and rearranging gives

$$s'(t_2' - t_1') = s(t_2 - t_1)$$

Since the following holds for the headway distance d_h:

$$d_h \leq s(t_2 - t_1)$$

then

$$d_h \leq s(t_2 - t_1) = s'(t_2' - t_1')$$

and, in fact, the same headway distance is achieved.

There are other ways to pre-define solutions for contingencies. As one more example, consider a flight that is informed that one of the lanes it is scheduled to traverse in its flight path is no longer available. It is possible that the reservation system actually reserves alternate paths to the destination so that the flight can choose one of these alternates if necessary. In this case, the flight will choose a reserved flight path that does not include the unavailable lane. As the flight passes turning points to take alternate routes, it can inform the UTM that releases those reservations and makes the lanes available for use by other flights.

7.1.4   Dynamic Contingency Mechanisms

Dynamic contingency handling allows more flexibility in responding to problems. It also generally requires more communication or sensing capabilities for the platforms (or agents) involved. The advent of 5G and beyond communication techniques is opportune for the development of dynamic contingency handling because of the increased bandwidth and the lower delay times. Consider first the case in which a UAS cannot follow its assigned schedule, and it communicates this to the UTM. At one extreme the UAS may be forced to land, and the UTM can dynamically create a landing lane suitable for the specific UAS and location. According to the type of UAS and the roads and buildings below it, a landing lane may be created to allow the UAS to land on a building top, or in the emergency lane of a road on the ground below it, or in an empty part of a nearby parking lot, etc. Moreover, it is also possible for the UTM to monitor cell phone usage in the landing area and choose a site with the fewest number of people.

Alternatively, if the flight is able to continue to its destination, but at a reduced speed, then it may be possible to allow other UAS to continue on their scheduled paths and for them to dynamically interact with the slowed flight in order to avoid it. For example, if emergency side lanes are available in the air lane network, then when a nominal UAS approaches the slowed flight, they can make sure that the minimal headway distance is maintained by detect and avoid measures (assuming the emergency air lane is close enough to the regular lane to warrant that). Negotiation of a passing maneuver can be handled by direct communication, intermediate communication through the UTM, or by using an established protocol. As part of our work on UAS dynamic tactical deconfliction we developed the *Closest Point of Approach Deconfliction Algorithm* (CPAD) [100]. CPAD was developed in the context of the Dynamic Data-Driven Applications Systems program from the Air Force Office of Scientific Research. This approach is described in detail below.

7.2   Real-Time Tactical Deconfliction

Each lane has a set of connected lanes with which it shares an endpoint. A flight in
a given lane is tactically deconflicted if there is no point in its trajectory along the
lane such that it is within distance d_S of any flight in a connected lane. This can be
efficiently checked using the Closest Point of Approach (CPA) algorithm as follows.
Let two lanes, $\mathcal{L}_1$ and $\mathcal{L}_2$, be defined by vectors $\bar{S}_1$ and $\bar{S}_2$, where $\bar{S}_1 \equiv \overrightarrow{P_1 P_2}$ and
$\bar{S}_2 \equiv \overrightarrow{Q_1 Q_2}$. The trajectories of flights f_1 and f_2 in lane $\mathcal{L}_1$ and $\mathcal{L}_2$, with velocities
$\bar{v}$ and $\bar{w}$, are defined as $\bar{P}(t) = \bar{P}_1 + t\bar{v}$ and $\bar{Q}(t) = \bar{Q}_1 + t\bar{w}$. Since the velocities
are $\bar{v} = \frac{s_1(\bar{P}_2 - \bar{P}_1)}{|P_2 - P_1|}$ and $\bar{w} = \frac{s_2(\bar{Q}_2 - \bar{Q}_1)}{|Q_2 - Q_1|}$, where s_1 and s_2 are the respective speeds of
f_1 and f_2, then the time, t_{min}, when the two flights are closest in their trajectories is

$$ t_{min} = \frac{-(\bar{P}_1 - \bar{Q}_1) \cdot (\bar{v} - \bar{w})}{|\bar{v} - \bar{w}|^2} $$

If t_{min} is found for $t \in [t_{current}, t_{min\_TOA}]$, where t_{min_TOA} is the minimum time
of arrival at the end of the lane for flights f_1 and f_2, then the minimum distance,
d_{min}, between the flights across these intervals is just $|\bar{P}(t_{min}) - \bar{Q}(t_{min})|$. If
$d_{min} < d_S$, then a conflict exists between the two flights. Figure 7.2 illustrates the
CPA method.

If a flight, f_1, has a conflict with flight f_2, then the two flights can be deconflicted
as follows:

Deconflict_Pair

while conflict(f_1, f_2)
    reduce speed, s_1, of f_1
    if $s_1 < s_{min}$
    then flight f_1 fails

Fig. 7.2 CPA algorithm: two
flights at closest points $P_{t_{min}}$
and $Q_{t_{min}}$

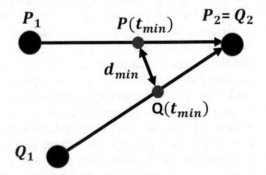

This allows the definition of the Closest Point of Approach Deconfliction (CPAD) algorithm:

Algorithm 1: Closest point of approach
1 ∀ active flight, f
2 **if** f enters a new lane
3   **OR** a neighboring flight has slowed
4   **OR** f has reduced speed on its own
5 **then** call Deconflict_Pair for all flights in neighboring lanes
6 **if** f has reduced speed
7 **then** f broadcasts this information.

7.2.1 Approximate Global Deconfliction Using CPAD

Global tactical deconfliction is achieved by having each UAS run the CPAD algorithm. CPAD does not guarantee strategic deconfliction (i.e., that no two flights get within distance d_S across the entire set of current flight plans); however, it does guarantee that no two flights are ever within distance d_S of each other at any time. The benefits of this approach include that that there is no centralized flight planning, no sharing of detailed flight info between USS, and robustness in the face of contingencies. The cost of the approach is that some flights may be forced to fail; however, this can be mitigated by choosing appropriate lane structure, controlling the number of flights, and eventually by dynamic flight route selection (currently the lane sequence is fixed). Certain communication requirements are imposed; however, the data shared between flights is essentially their telemetry data, which the FAA-NASA UTM requires broadcasting anyway.

7.3 Experiments

A discrete event simulation is run, which allows specification of the simulation time interval, $[0, t_{max}]$, and the number of flights, n_f. One unit distance corresponds to 50 ft, and one unit time corresponds to 10 seconds. Two maximum speeds are considered: 5 and 9, which correspond to about 17 and 31 mph, respectively. Each flight has randomly selected launch and land sites, as well as a random desired launch time. A fixed 3×4 grid of lanes at altitude 10 units is serviced by 10 launch lanes and 10 land lanes (see Fig. 7.1).

When a flight plan is created for a flight, it consists of a sequence of lanes and for each a specific Time of Departure (TOD: departs entry point to lane) and Time of Arrival (TOA: arrives at exit point of lane). The next event is just the flight with the earliest TOA in its current lane, unless it has not yet launched in which case it is the current launch time. The launch times of the flights are uniformly distributed across the simulation time interval. Note that if a flight cannot launch at its desired launch time due to conflicts in the launch lane, then it is rescheduled to a later time (with fixed delay). Once an event is selected, all flights are advanced according to their respective speeds in their current lanes. Next, the flights are deconflicted.

We consider two aspects for study: (1) maximum simulated time (set to 100 and 200 units), and (2) maximum UAS speed (set to 5 and 9 units distance per unit time). These correspond to about 17 and 33 minutes, and 17 and 31 mph, respectively. The number of flights is chosen to equal the maximum time since this represents on average one launch per launch site every 50 seconds. Given a max time, UAS max speed, and number of flights, the simulation is run using the CPAD algorithm. Table 7.1 gives the data for five representative runs, as well as the means. As can be seen, these results indicate that the CPAD algorithm works well in these scenarios with only one flight failure in all of the experiments (3000 flights overall). Moreover, the average speed is quite near the maximum allowed speed, and there are very few

Table 7.1 Delays and failures in experimental simulations

t_{max}	n_f	s_{max}	Wait	Fly	Done	Fail	Avg speed	Delays
100	100	5	1	18	81	0	4.98	2
			2	12	86	0	4.98	2
			0	15	85	0	4.99	1
			0	11	89	0	4.98	2
			1	18	81	0	4.96	4
Means			0.8	14.8	84.4	0	4.98	2.2
100	100	9	0	11	89	0	8.98	1
			1	8	91	0	8.94	2
			0	12	88	0	8.99	0
			0	6	94	0	8.99	0
			0	11	88	1	8.98	0
Means			0.2	9.6	90	0.2	8.98	0.6
200	200	5	0	14	186	0	4.96	6
			0	11	189	0	4.97	8
			0	17	183	0	4.98	6
			1	13	186	0	4.99	10
			0	6	194	0	4.96	9
Means			0.2	12.2	187.6	0	4.97	8.6
200	200	9	0	7	193	0	8.96	4
			1	6	193	0	8.97	2
			0	8	192	0	8.97	4
			0	7	193	0	8.98	3
			0	4	196	0	8.97	2
Means			0.2	6.4	193.4	0	8.97	3

delays (68 out of 3000). The most critical parameter for algorithm performance is the maximum speed of the UAS. Other trends revealed in the data include that the longer the time period, the more flights complete their mission, and the fewer flights are delayed or in the air (on average).

7.4 Conclusions and Future Work

The lane-based approach provides a viable model for large-scale urban air traffic, and CPAD closes the symbiotic DDDAS feedback loop to update the model based on measurements and communication as required by the model. The results here lay the foundation for a further study into the role of DDDAS in large-scale unmanned traffic management. System designers must consider the impact of airspace structure on information flow as well as the accessibility of the network (measured as delay). This demonstrates the importance of considering the structure of the discretization of the configuration space and how a real-time dynamic flight deconfliction algorithm can operate under strong assumptions about the space/time structure of the environment. Future issues to be explored include: (1) a broader set of experiments will be run to study the role of the number of lanes, the distribution of flights over lanes, etc., as well as a sensitivity analysis of the experimental parameters, (2) flights are assigned a complete sequence of lanes in this study, but we intend to explore the application of the software defined networking paradigm to dynamically select the lane sequence, (3) the structural properties of the airway network also play a role in facilitating flight deconfliction, and those parameters will be studied, (4) experiments will be conducted on realistic airways scenarios; e.g., the Utah Department of Transportation is exploring the use of the lane-based approach in Utah, where the airways are located above roadways, and (5) CPAD imposes communication requirements on the aircraft, and this aspect will also be studied in terms of the likelihood of failure to communicate correctly and its impact on deconfliction.

Chapter 8
Agent Based Modeling and Simulation

8.1 Introduction

The ability to statically analyze UAS traffic management systems (UTM) is hampered by the dimensionality of possible behaviors. Individual agents can act in different ways if their algorithms are not standardized, and even then there are contingencies that can thwart the best plans available. The collection of individual agents and their behaviors form the system, and one way to model the collective behavior is through Agent-Based Modeling and Simulation (ABMS). Various Agent-Based Modeling and Simulation software frameworks exist (see [24, 66, 77]); however, a specific framework was developed here in order to find a set of symbiotic UAS behaviors and UTM policies. The framework is instrumented to allow measurement of crucial features, including local statistics and flow metrics, contingencies (and if possible their causes), and higher-level system features and emergent behaviors. Our previous work on the BRECCIA system [95, 97] included a BDI-agent-based framework built from a Java library called Jason [21]. However, the ability to rapid-prototype different models of communication, or create complex agents, is inhibited by having to switch between Java and the Jason domain-specific language. This chapter describes the ABMS approach to analysis for UTM systems.

8.2 Lane Systems and Sensitivity

Once the airway network is defined, a lane-based strategic deconfliction algorithm is required to schedule flights into the lanes so as to maintain the required minimum separation at all times during flight; this assumes that every flight follows its approved flight plan. The *Lane-Based Strategic Deconfliction* (LBSD)[1] algorithm is given, which allows computationally efficient scheduling. It is shown that this algorithm is $\mathcal{O}(k^2)$, where k is the number of flights in the lane sequence of the proposed flight during its flight.

Alternatively, in-flight planning arises due to contingencies, i.e., possible future events, usually causing problems or making further plans and arrangements necessary. Contingency handling may occur at different time-horizons and require different mechanisms, for example tactical (sensor-based) deconfliction. For these scenarios the *Closest Point of Approach* (CPA) algorithm is defined so that UAS can exploit the lane structure to continue their flights while avoiding collisions. This protocol may be based on either individual UAS sensor data or on local inter-UAS communication.

8.3   Lane Systems and Robustness

The layout of the lane system can also have significant effects on the behavior of the system. A common refrain among air mobility enthusiasts is that the ability to travel point-to-point in a straight line, Fig. 8.1 for example, should be maintained and decreases the desirability of structured airspaces. However, a system of agents performing individually optimal trajectories in an unstructured airspace is unlikely to produce an efficient system. This is true in the case where agents can make decisions dynamically based on system-wide conditions, for example, Braess' paradox demonstrates where additional route options can result in an increase in travel time [35]. This also appears to be true when considering conflict counts for a simple cell-based deconfliction experiment (point-to-point flights deconflicted using ground-delay, we call the FAA-NASA approach). Figure 8.2 shows histograms for cell traversals (how many times a flight crossed a cell) and intersections (how many flight paths intersect) for an experiment with 1000 UAS flying point-to-point in an unstructured airspace with uniformly distributed land and launch sites. These graphs show an increased density of conflicts focused in the center of the area of interest.

An ABMS framework that simulates AAM should model the roles and responsibilities of the real UTM framework. However, the separation of responsibilities between the USS (or PSU) and the operator can be merged for the purposes of analyzing the resulting traffic and system behavior, since this separation mainly serves regulatory requirements. Therefore a reasonable organization for an ABMS includes the following organization:

---

[1] ©[2022] IEEE. Reprinted, with permission, from [IEEE-T Intelligent Transportation Systems, "Lane-Based Large-Scale UAS Traffic Management," David Sacharny, Thomas C. Henderson and Vista Marston, 2022, Print ISSN: 1524–9050, Online ISSN: 1558–0016, Digital Object Identifier: 10.1109/TITS.2022.3160378].

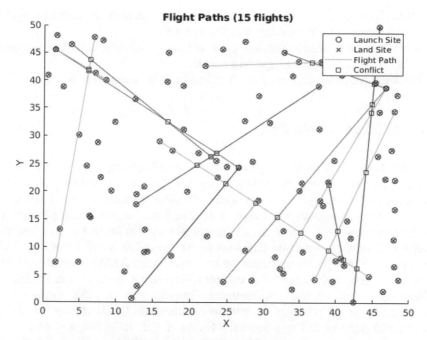

Fig. 8.1 Sample straight line paths between launch and land vertices

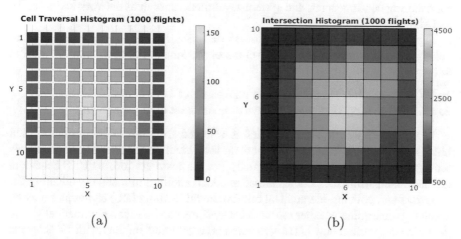

Fig. 8.2 Unstructured airspace density and path intersections. (**a**) Cell traversal counts. (**b**) Path intersection counts

- LBSD: This class encapsulates the Lane-Based Strategic Deconfliction (LBSD) Supplemental Data Service Provider, responsible for approving reservations into the airspace.
- ATOC: An instance of this class represents an Air Traffic Operations Center, providing users with visualization capabilities.

- UAS: Unmanned Aerial System is a mobile agent that operates within the UTM.
- KB: A database for knowledge storage and access.
- SIM: This class encapsulates all the simulation functions, including mocking GPS and radar sensors and updating agents.
- RADAR: An encapsulation of a reduced-order radar sensor model.

8.4   ABMS Optimization

The complexity of determining good plans for handling contingencies is a central problem for large-scale autonomous systems. Different adaptations of the lane-based approach can represent the majority of proposed airspace structures, from the least structured (every UAS creates its own lane, i.e., the free-flight model), to the most structured, where regulators completely determine the layout of the airspace. How these structural decisions affect the behaviors of individual UAS, and ultimately the system dynamics are captured by the ABMS approach. ABMS is ideally suited to the analysis of complex, large-scale systems with interacting heterogeneous agents, and can incorporate cognitive models that mediate how individual agents perceive and react to the environment. By alternately optimizing individual behaviors and system-level policies, a symbiotic design is found that balances individual preferences, which include the ability to make good decisions for handling contingencies, and system-level metrics, such as network accessibility and efficiency.

Figure 8.3 shows the iterative process that uses ABMS to assess policies and behaviors to demonstrate the effectiveness of the lane-based approach. The goal is to optimize:

1. **UTM Policies**: UTM policies over a given set of UAS behaviors.
2. **UAS Behaviors**: UAS agent behaviors given a set of UTM policies.

In each iteration, a set of policies is selected for a given environment, then UAS behaviors are adapted to improve individual and system-level measures (compromises are explicitly documented). For the next iteration, the UAS behaviors are held fixed while the policies are adapted. In addition to manually adapting the behaviors and policies, hierarchical reinforcement learning (RL) techniques may be applied to determine whether more robust systems can be found automatically. The goal is to understand how UTM structure and constraints interact with UAS agents to produce emergent behaviors that impact system performance. In particular, it is essential to avoid cascading non-conforming flights arising from contingencies.

8.4.1   Contingency Handling as a Measure of Effectiveness

Contingencies are important because they represent the safety and logistics issues that arise in real-world systems, serving major design and operational constraints. Contingencies also represent a considerable computational challenge, since the

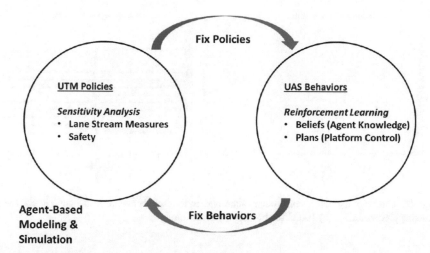

Fig. 8.3 Agent Based Modeling Framework for Learning UTM Policies and UAS Behaviors. The UTM policies are set, and then UAS agents learn both at the BDI cognitive level (beliefs) as well as at the platform control level (plans). UAS behaviors are fixed and UTM Policies are then optimized with respect to global lane stream properties as well as safety measures

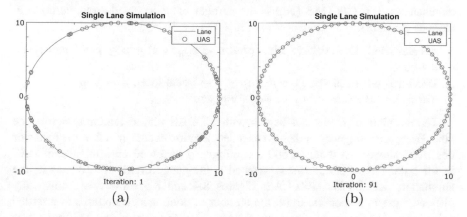

Fig. 8.4 Single-lane contingency effects. (**a**) UAS after one iteration. (**b**) UAS after 91 iterations

representation and planning for all possible effects is generally intractable. In fact, contingencies are the primary impediment to the large-scale integration of autonomous systems. Therefore, to demonstrate that the lane-based approach is an effective organizational strategy for UTM, it must be shown that the lane-based approach provides a low-complexity foundation for contingency analysis and mitigation.

As an example of how the lane-based approach supports contingency analysis, consider the single-lane example in Fig. 8.4 (adapted from [64]). In this model, the lane is a one-dimensional curve represented by an array of length L. Each element

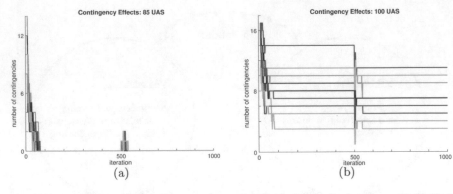

Fig. 8.5 Contingency effects on system state and individual behavior in a lane. (**a**) 85 UAS showing predictability. (**b**) 100 UAS showing unpredictability

of the array can be in one of seven states: It may be empty, or occupied by a UAS having an integer speed of zero to five. The speed value represents the number of array elements that UAS moves forward in the next step of the simulation. The behavior of each UAS is defined as follows, calculated at each step of the simulation simultaneously for all UAS (gap is the number of unoccupied array elements in front of the UAS):

1. **Acceleration:** Each vehicle with speed $v < v_{max}$ and $gap \geq v + 1$ gets speed $v \leftarrow v + 1$.
2. **Deceleration:** Each vehicle with $gap \leq v - 1$ gets speed $v \leftarrow gap$.
3. **Move:** Each vehicle moves forward v elements.

The simulation in Fig. 8.4 begins with 85 UAS placed randomly across the lane. Here, a contingency is defined as a deceleration event (item 2 in the behavior described above). At iteration 500 a contingency is forced on a number of UAS in the lane. Figure 8.5 shows aggregated contingency events for 10 runs of the simulation with 85 and 100 UAS. Figures 8.4 and 8.5 show two dramatically different system responses: In (a) the number of contingencies returns to a settled value before the forced contingency, while in (b) each run produces a different, and hence unpredictable,[2] outcome.

The lane-based approach provides a single control variable to account for these two scenarios: the density of UAS in the lane structure. A more complex individual behavior could potentially guard against such unpredictable effects, but with this approach the trade-offs are made explicit for the policy maker.

---

[2] As noted by Nagel and Rasmussen [64], this model can be treated analytically [103], but the analytical results are "more difficult to obtain" than measurements from simulation. A result that would support this thesis would show that free-flight systems, the least structured of airspace designs proposed, are more difficult to analyze than lane-based systems and therefore less ideal for contingency handling.

8.4.2 UAS Behaviors for Contingency Handling

Consider a scenario in which ground infrastructure supporting UAS communications is disrupted during normal operations. Currently, the published protocol for handling this contingency is to fly back to base if communications cannot be re-established within a given amount of time [10]. Since this is a pre-defined policy, it is worth considering whether such a policy is robust. For example, depending on how many UAS communications are disrupted, the number of conflicts that result from the simultaneous replanning of multiple agents may have negative cascading effects [52]. As the complexity of the UTM system increases, it becomes harder for experts to enumerate all the failure modes and effects; assigning liability and performing post-failure diagnoses will also be difficult. Table 8.1 gives the set of contingencies considered here, the contingency response, and the required communication capabilities for the response.

8.4.2.1 The Role of Cognition in Contingency Handling

To develop robust policies for UAS in the UTM system, one can frame the problem in a similar manner as designers did for ACAS X. Another method is to use a hierarchical decomposition of states and actions, and a semi-Markov decision process model for state-transitions [5, 9, 87]. A benefit of the lane-based

Table 8.1 Contingency table; V2V (Vehicle to Vehicle V2V), V2G (Vehicle to Ground), and V2X (Vehicle to Everywhere)

Contingency	Responses	Communication requirements
UAS Nav loss	Get position from other UAS	V2V, V2G
	UAS operator manual control	V2G
	Tactical landing	V2X, V2G
UAS speed loss	Move to emergency lane	V2V, V2G
	Alternate flight path	V2G
	Tactical landing	V2X, V2G
UAS energy loss	Plan alternate route to destination	V2V, V2G
	Alternate destination	V2V, V2G
	Nearest destination	V2V, V2G
	Tactical landing	V2X, V2G
UAS Comms loss	Wait in emergency lane	
	Tactical landing	
UAS falling	Deploy parachute	V2V, V2G
Lane obstacles	Tactical avoidance	V2V, V2G
	Move to emergency lane	V2V, V2G
Lane closure	Alternate route	V2G
	Create lanes	

system is that a concise description of surrounding vehicles, their intentions, and locations is provided by the space-time lane diagram (lanes, reservations, headways, and speeds). The STLD representation of system-level state can be fed to each UAS agent for the purpose of contingency planning, potentially leading to higher preference behaviors. For example, in a communications-loss scenario, a UAS may decide to seek a different emergency landing pad if it knows that other agents in the lane are likely to execute the same contingency plan. Another possibility is that two UAS with communications loss continue past the disruption because other UAS farther along appear to be functioning nominally, especially if a goal encoding relay-communications were included. Such a scenario-specific policy would be difficult to program explicitly, but such robust solutions are feasible through policy optimization.

A hierarchical reinforcement learning technique can automatically determine actions, as well as policy. More recent developments have shown promise in learning games by creating abstractions of the actions available to agents. Both the Option-Critic Architecture [9] and FeUdal Networks [120] take advantage of semi-Markov processes to model different time-scales in the hierarchy, but FeUdal networks benefit from an explicit representation of a high-level goal. The BDI architecture is a convenient way to structure and analyze agent behavior, and includes a hierarchical decomposition of actions in the form of plans. The different decompositions of actions for operators, USS, and airspace controllers in the UTM system are compared in Sect. 8.5.1 on Experiments. For example, two competing high-level goals in the UTM system are to maximize headway and lane density. Actions available to agents include *Replan*, *Land*, *Cruise*, etc. Additionally, since NASA has published the required application programming interfaces (APIs) for USS, the possibility of automatically generating low-level actions from these APIs is also available.

8.4.3   UTM Policies for Contingency Handling

An intelligent airspace provides a robust and redundant basis for contingency handling, such as the one-way lane structure, i.e., by eliminating, in the nominal case, any necessity for UAS to coördinate through intersections. However, the lane is more than a simple curve in 3D Euclidean space. Some properties that must be specified include the lane boundary surface (e.g., a tube), minimum and maximum speeds, geometry (straight, curved), length, etc. Another issue to be considered is whether regular lanes between two intersections going in opposite directions should be aligned left-right at the same altitude (as standard ground roads) or one above the other.

In addition, lane placement should be conditioned on several crucial operational aspects, some of which include: (1) *ground structures*: buildings, towers, etc., (2) *population density* that dynamically varies with time, (3) *roadways on the ground*: these may be avoided or followed, (4) *communications ground stations*: that may also provide network repair if ground stations fail, (5) *landing sites*: availability of emergency landing sites, and (6) *monitoring infrastructure*: radar, microphones, etc. Some of these choices are not dichotomous. For instance, placing a lane above a road will help to reduce aerial obstacles and improve communications as UAS can use the cellular wireless infrastructure that has full coverage in most of the roads and communicate with ground vehicles and road side units below them using the unlicensed dedicated short-range communications (DSRC) channels [55] that might lead to better safety; however, such a lane assignment increases the population density below the flight, and thus, the risk of human injury.

There are four basic types of lanes that the trajectory of a UAS can encounter: (1) *launching/landing*: allow ingress, egress to/from airspace, (2) *regular*: allow travel through airspace (intersection to intersection), (3) *roundabout*: provide way through intersection, and (4) *ramp*: transition between regular and roundabout lanes. There are several parameters that constrain the placement and connectivity of lanes. For example, a regular lane may be at a different altitude from roundabouts, and therefore ramp lane placement must accommodate the individual lane parameters of each other type.

8.5 ABMS Test: FAA vs. Lane-Based Approach

To schedule a flight, launch and land sites are selected, as well as a sequence of lanes going from each one to the other, along with a desired speed, and a launch time window. The set of lanes may be selected however desired; for example, to minimize distance or weather constraints, or other relevant factors. The launch time window gives the earliest and latest possible launch times (line 1 of LBSD algorithm). Lanes are scheduled individually by flights, and every new flight must respect the headway distance not only in each lane, but also when moving from one lane to another (i.e., with respect to all merging or diverging lanes). The Lane-Based Strategic Deconfliction algorithm used in the following experiments was described in Chap. 4, Algorithm 1: Label Method.

A requirement of the Label Method is that a complete database of flight reservations must be maintained and used by the algorithm; however, this will generally be required by the flight authorities anyway to allow informed monitoring of airspace usage. The original idea of the FAA was to allow a decentralized approach where each USS maintained its own flight info and shared as necessary; the drawback of this is that if any USS fails, the system fails, and there is the possibility of semantic mismatch in terms of trajectory definition (e.g., meters vs. feet).

The Space-Time Lane Diagram (STLD) that is enabled by the lane-based approach also provides a straightforward way to visualize the traffic through a lane for monitoring UTM operations. Figure 8.6 shows a set of planned flights through a lane, where reservations represent a reduced-order model (speed and headway) for the actual or planned trajectory. And their trajectories reflect the accelerations

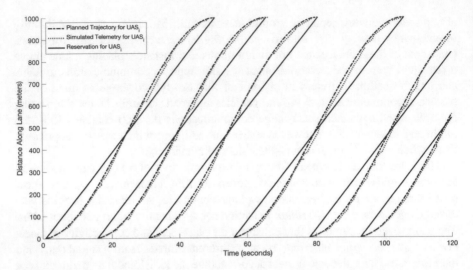

Fig. 8.6 Lane diagram for a single lane, showing six flight reservations, planned trajectories, and simulated telemetry

necessary to turn between lanes. Lanes also allow real-time comparison of the UAS' planned flight path and the actual trajectories (e.g., provided by telemetry data).

8.5.1   Experiments to Determine Parameter Impact on Scheduling Algorithms

In a complicated system like a UTM, analytic solutions may not exist, and therefore, simulations are used to explore UTM performance with respect to parameters of interest. The experiments performed here are designed to allow both inter-UTM (e.g., LBSD vs. FNSD) and intra-UTM (e.g., grid vs. Delaunay) structural analysis, as well as a cursory system/behavioral analysis (relating the agents flexibility in scheduling to the overall system performance). The parameters studied here include:

1. *Launch Frequency* (flights per hour): Comparable to an arrival rate of flights into the system [values: 100 and 1000].
2. *UAS Speed* (m/s): Average UAS speed through lane [values: 5, 10, 15].
3. *Headway Distance* (m): Minimum distance allowed between UAS [values: 5, 10, 30].
4. *Flex Time* (sec): Interval of possible launch times for flight [values: 0, 300, 1800].

The simulation covers an area of 5 square km (roughly the size of the Salt Lake Valley) with the FAA cells spaced as a 10x10 cell structure. The LBSD grid was chosen to correspond to this as an 11x11 node grid. The 121 launch (land) sites are located near the ground node points in both layouts. The Delaunay networks are generated with the same number of nodes, but they are distributed randomly (sampled from uniform distribution) in the given area. Road-based networks include an area over San Francisco and area over Salt Lake City. Ten simulation trials were run for each of the 54 parameter combinations (note that for the Delaunay networks an additional ten trials were run for each due to the random nature of the node locations). The simulation period was set to 4 hours simulated time. The FAA flights are up, over, and down trajectories scheduled between randomly selected launch and land sites; the flight altitude was randomly assigned between the min and max altitudes of the LBSD network. For both UTM methods, given the flight frequency, a random set of desired flight times are generated, which are uniformly spread across the total simulation time.

Figures 8.7 and 8.8 show the mean statistics for launch frequency of 100 flights/hour and 1000 flights/hour, respectively. The means of the maxima over all

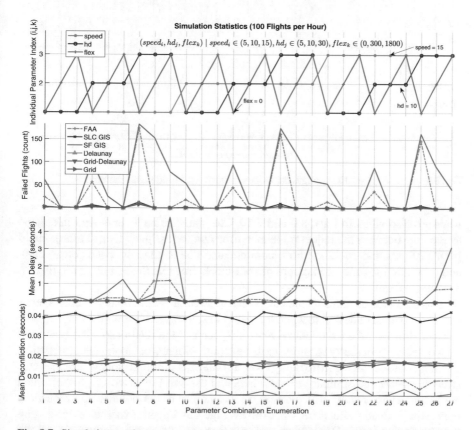

Fig. 8.7 Simulation results: averages for launch frequency of 100 flights per hour. The upper row describes the parameter combination enumeration in the lower three rows, which give the mean number of failed fights, mean delay, and mean deconfliction for those combinations of parameters

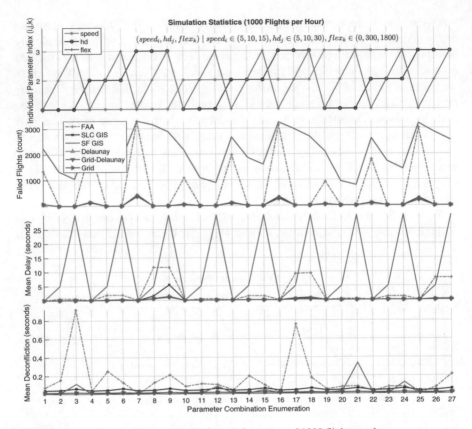

Fig. 8.8 Simulation results: averages for launch frequency of 1000 flights per hour

trials are also given in Figs. 8.9 and 8.10. The statistics include: delay (calculated as the time between the requested launch time and the assigned launch time), failed flights (flights that could not be accommodated due to time or space constraints), and deconfliction time (the amount of wall-clock time that the computer required to schedule a flight).

This data indicates that all six categories of structures have response characteristics that are most undesirable when the flex is low, the speed is low, and the headway is high. However, the unstructured FAA airspace and the road-based San Francisco networks are particularly sensitive to these inputs with respect to the mean statistics. The max statistics in regard to delay show a somewhat different story where the FAA structure responded similarly to the others and the San Francisco graph performed the worst. These results indicate that small changes in the policies and behaviors may have dramatic effects on what the average UAS agent experiences accessing the unstructured (FNSD) airspace and complex road networks. Conversely, all the structured airspaces had relatively subdued effects related to these inputs (note that Salt Lake City has a grid-like road system).

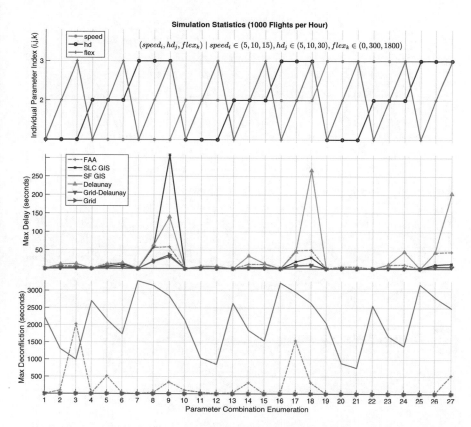

Fig. 8.9 Simulation results: maxes for launch frequency of 1000 flights per hour

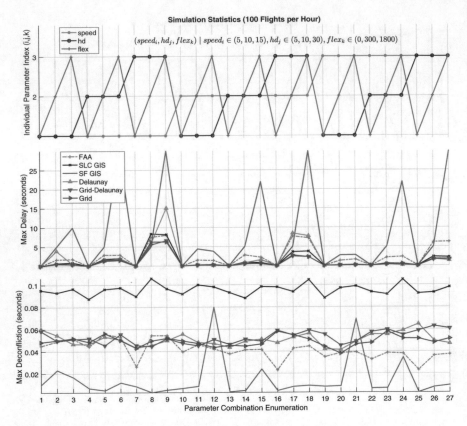

Fig. 8.10 Simulation results: maxes for launch frequency of 100 flights per hour

Chapter 9
Strategic Deployment of Drone Centers and Fleet Size Planning for Drone Delivery

9.1 Introduction

The parameters of a specific UTM application are important not only for considering the feasibility of Unmanned Aircraft Systems (UAS) package delivery within the state but also for determining the impact and ultimately the efficient operation of drone delivery. UAAMS allows the Utah Department of Transportation (UDOT) to assess different assumptions of the model and run "what-if" scenarios by generating animation of the optimized airspace network. The platform provides the state with more clarity about the energy impacts of large-scale drone delivery, as well as a viable airspace network. The tool can further inform the UDOT Division of Aeronautics to develop policies and negotiate with industry stakeholders.

Through this UTRAC research project, the implementation and deployment of the UAAMS was successful and will be available for continuing research in the development of advanced air mobility in the state of Utah. This type of tool is critical for research in this area, as it incorporates the latest software and infrastructure development techniques available. Also shown was the viability of considering the large-scale impacts (e.g., environmental) of advanced air mobility on specific communities by using micro-simulation technology. In contrast to micro-simulations performed for human-controlled ground and air traffic, in this case a model for the autonomous agents has the potential to be exact, since their algorithms must be documented. Future improvements will consider optimizations to the object function in the optimization procedure—this would allow a more thorough exploration of the possible configurations of vehicle and vertiport parameters.

D. Sacharny, T. C. Henderson, *Lane-Based Unmanned Aircraft Systems Traffic Management*, Unmanned System Technologies, https://doi.org/10.1007/978-3-030-98574-5_9

9.1.1 Problem Statement

In a 2017 report by the RAND corporation [62], analytical methods for calculating the total energy consumed by a mix of delivery trucks and drones were developed and shown to be highly dependent on the layout of distribution centers as well as distance traveled by delivery vehicles. This suggests that the city layout, i.e., street connectivity and other network parameters, are important considerations for energy-conscious policies. While industry stakeholders must determine the market viability of drone delivery, they are not required to calculate the external and indirect costs that may be associated with this burgeoning industry. The web-based platform developed for this report, called the Utah Advanced Air Mobility Simulator (UAAMS), enables researchers, planners, and practitioners to record and update assumptions about the distribution of vertiports, traffic, population, and other requirements that may affect the operation of the transportation network. These parameters are important not only for considering the feasibility of Unmanned Aircraft Systems (UAS) package delivery within the state but also for determining the impact and ultimately the efficient operation of these new transportation technologies. Furthermore, additional analysis and what-if scenarios may be developed using this simulator. Example Jupyter notebooks (python) are provided to help guide development; however, users are not limited to any particular programming language. To facilitate the iterative process needed for the development of Advanced Air Mobility in the State of Utah, UAAMS is a web-based software and can be accessed from any web browser. Additionally, the simulator is delivered with the open-source web-map server, GeoServer (https://geoserver.org), which contains the geospatial data used or generated by the simulations. This enables multiple agencies within the state, who often use Geographic Information Systems (e.g., ArcGIS), to communicate planning efforts and incorporate data from, or provide data to, these simulations. Figure 9.1 shows the main features of the software developed for this project:

1. Simulator Form: The simulator form enables the client to enter assumptions about the vehicles and the environment, and then execute the simulation.
2. 3D Map Interface: A viewer for map and simulation data.
3. File System Explorer: Input and output files for simulations (e.g., vertiport locations and result figures) may be stored in the filesystem. Additionally, all the source codes for simulations are accessible through this file system explorer.
4. Simulation Output Viewer: Simulations produce a number of figures that can be viewed here.

An initial simulation implementation was developed for this report to demonstrate the workflow for running simulations, as well as developing new ones, and is described in the sections that follow. The UAAMS is deployed to US data centers on Google Cloud and is accessible by anyone with authorized credentials by visiting the URL https://utrac.georq.io.

File System Explorer

Simulator Form

3D Map Interface

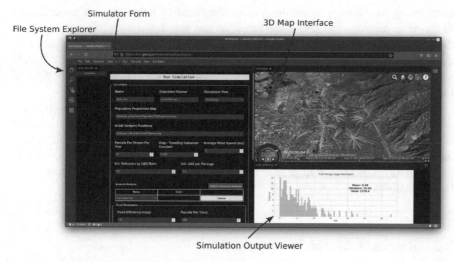

Simulation Output Viewer

Fig. 9.1 The Utah advanced air mobility simulator (UAAMS)

9.1.2   Objectives

The motivation for this work is to provide regulators, policymakers, and industry stakeholders with a data-driven framework for assessing the energy costs and trade-offs of large-scale drone delivery in the state of Utah. The primary objective is to produce a web-based platform that takes inputs of state-wide road network, the total number of (drone-deliverable) packages to deliver on a given day and their destinations, and energy and cost assumptions per vehicle, and produces a state-wide airspace network, delivery schedule, and truck/drone fleet mix. A secondary goal is to optimize the network to ensure that drones are strategically deconflicted as required by FAA/NASA and the total energy over that day is minimized. Overall, this program will provide the state with more clarity about the energy impacts of large-scale drone delivery, as well as a viable airspace network.

9.1.3   Scope

This research involves three major components: data collection, optimization model development, and web-based platform development. We first gathered data on the state-wide roadway network, population data (year 2025), Traffic Analysis Zone (TAZ) boundary, and post office locations. The dataset enabled the creation of delivery zones to simulate package delivery coverage area for each drone center. In addition, the air delivery network was created by lifting the virtual highway network into the sky. This includes generating airways, deconflicting zones for Utah considering the delivery scheduling, launching/landing lanes, etc. An optimization model was developed to determine the schedule and deployment of drones within the study area. The developed system model with the optimization component was then implemented onto a web-based platform where people can assess different assumptions of the model and run what-if scenarios by generating animation of the optimized airspace network.

The rest of the discussion is structured as follows. Section 9.2 summarizes the research methods. Section 9.3 illustrates the web-based platform along with a case study for a medium-sized simulation with 33 TAZs. Section 9.4 presents the results and findings, and outlines the lessons learned for follow-up research. Appendix D of the UDOT Technical Report [93] includes the user guide of the developed web-based platform.

9.2   Research Methods

9.2.1   Overview

Large-scale drone delivery is on the horizon nationwide, as it has the potential to decrease pollution and help alleviate road congestion. Up until now, industry is mainly concerned with the market viability of drone deployment, and very little attention has been paid to external costs such as energy trade-off. The benefits of drone deployment, however, are largely dependent upon layout of distribution centers and distance traveled. To this end, we employ a data-driven approach to strategically replace ground-based delivery networks with air-based drone deployment. In this project, we structured the airspace by regulating and treating it as lanes, and optimized drone dispatch. Figure 9.2 shows the methodological framework of our research.

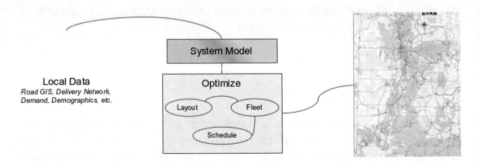

Fig. 9.2 Methodology framework of proposed drone network optimization

9.2.2 Data Source

The main data sources utilized to be fed into the airspace network construction consist of four parts:

1. Population projection: This is retrieved from the Utah Geospatial Resource Center (UGRC) [116] and is collected at the TAZ level. Year 2025 data was retrieved as the demand input. The population projection was further converted into parcel demand estimation, where on average 21 parcels/capita/year was assumed initially [42].
2. Post office locations: This is retrieved from UGRC [118] to serve as potential truck delivery centers. Figure 9.3 shows the overlay of post office locations with the TAZ boundary.
3. Utah road network: This is retrieved from UGRC [117] and the layout is used to create the elevated virtual airways in the sky for the drone fleet.
4. Utah building footprints and addresses: This is retrieved from UGRC [115] and provides the locations of possible drone deliveries for demand modeling.

9.2.3 Optimization Setup

The optimization procedure considers possible locations for vertiports to answer the question: Does there exist a set of vertiport locations that minimizes the total energy requirements? Since the spatial distribution of vertiports affects which addresses are served by them, the distances traveled and energy consumed by UAS are affected by their placement. Other parameters, such as vehicle characteristics, the proportion of packages that are delivered by UAS, and the routes of UAS and trucks have important effects, but were not considered in this procedure. To gain some intuition about how the locations of drone ports can affect the deliveries of packages, consider the relative locations of a drone center, truck center, and two TAZs in Fig. 9.4. At the start of the optimization, the drone center and the truck center are co-located. Package deliveries are scheduled according to the procedure that assigns aircraft to addresses that are within range and capabilities (according to the simulation parameters), and then fills the resulting demand with truck deliveries. For more efficient simulation, the truck distances and energy are calculated by solving a probabilistic traveling salesman problem [62]. In contrast to the regular shaped cells used for this calculation in [62], this simulation calculates the resulting convex hull that encompasses the truck-assigned demand.

At each iteration of the optimization, the drone centers are moved to a new location, the simulation is re-run, and the resulting total energy is calculated. The optimization is complete when further updates to the locations of the drone centers result in an increase in the total energy consumption. Several methods for optimizing vertiport locations were considered for this project, with the settled approach being

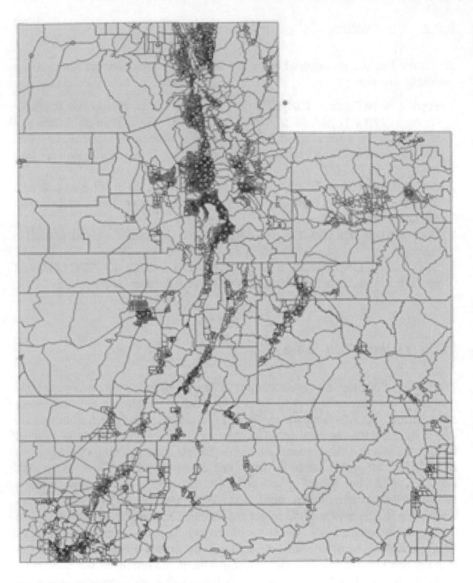

Fig. 9.3 TAZs with post office locations

the Covariance-Matrix Adaptation Evolution Strategy (CMAES) [28]. The objective was to minimize the total energy required by both UAS and trucks to deliver a set of packages in an area. Only the locations of the vertiports were considered variables in the optimization, while the job mix, number of vertiports, and vehicle parameters were held constant. Therefore, the number of variables in the optimization equals the number of vertiports under consideration. Figure 9.5 shows a plot of a sample optimization run for a single TAZ.

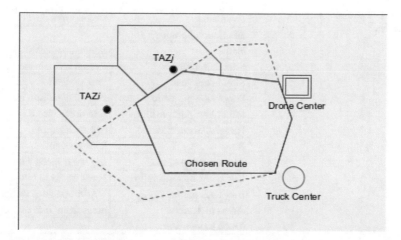

Fig. 9.4 A diagram showing how delivery routes (in red) are influenced by the placement of drone centers (i.e. vertiports) and truck centers relative to the travel analysis zones (TAZ)

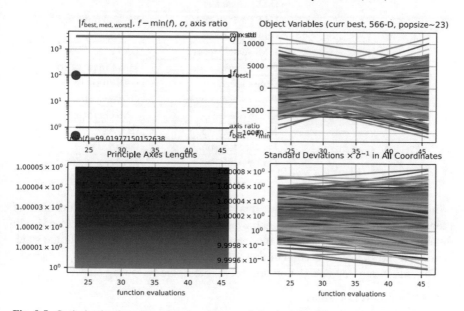

Fig. 9.5 Optimization iterations showing divergent behavior

9.3   Web-Based Platform

9.3.1   Overview

A simulation that extends what was described in the RAND report is provided along with UAAMS. This simulation incorporates all the parameters listed in Table 9.1.

Table 9.1 The simulation parameters

Form field	Description
Simulation parameters	
Name	Simulation name
Simulation runner	Python script name
Simulation year	Pop. projection year
Population projections map	shapefile/geojson file
Initial Vertiport positions	post office locs.K
Parcels/Person/Year	avg # packages
Ktsp	TSP constant
Average wind speed	avg wind speed felt
Init. deliv. by UAS ratio	From trucks to UAS
Init. UAS per package	UAS/package at vertiport.
Areas to Analyze	areas from TAZ dataset
Truck parameters	
Truck efficiency	Truck efficiency mpg
Parcels per truck	Truck package load
UAS parameters	
UAS payload mass	Payload mass (kg)
Lift-to-drag ratio	Of UAS
UAS vehicle mass	UAS mass (kg)
Power transfer efficiency	Motor to propeller
Cruise speed	UAS cruise speed
Power consumption	Of electronics
Climb rate	UAS climb rate
Maximum range	Of UAS
Package load time	Onto a UAS
Cruise Altitude	UAS cruise altitude (m)

In contrast to the RAND report, this simulator considers the real locations of addresses in Utah [118], as well as the projected population densities in order to simulate projected demand that is more accurately distributed throughout the state. Figure 9.6 shows a cross-section of these datasets.

The simulation implementation progresses along the following steps:

- Load Vertiport Locations
- Load Truck Depot Locations
- Load Population Projections (TAZ Zones)
- Load Building Locations
- Initialize UAS Model
- Filter TAZ Based on Requested City Areas
- Create Demand (Parcel Requests for a Day)
- Source Parcel Requests to Nearest Vertiport
- Calculated Parcel Requests within UAS range and Capability
- Calculate Truck/UAS Job Mix

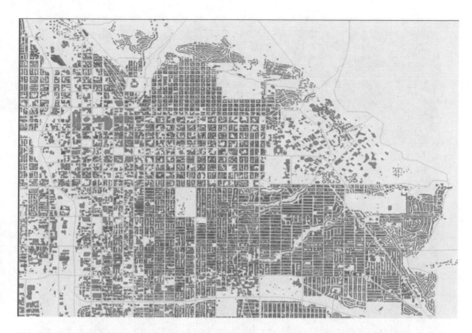

Fig. 9.6 Utah Buildings dataset overlaid on population projections dataset

- Generate UAS Trajectories
- Estimate UAS Round-Trip Times
- Estimate UAS Energy Requirements
- Estimate Truck Energy Requirements
- Generate Animation
- Generate Results

9.3.2   Case Study

A medium-sized simulation was executed that included 33 TAZs as a case study to demonstrate the web-based platform.

9.3.2.1   Simulation Form

A simulation form is displayed whenever a file with the suffix .sim is clicked in the workspace explorer. As this form is edited by the user, the parameters are written to the .sim file (the raw json format can be viewed by opening the file in the code editor as shown in Fig. 9.7). When the Run Simulation button is clicked at the top of the form, the python program defined in the Simulation Runner field is executed and the .sim file is provided as an argument to the program.

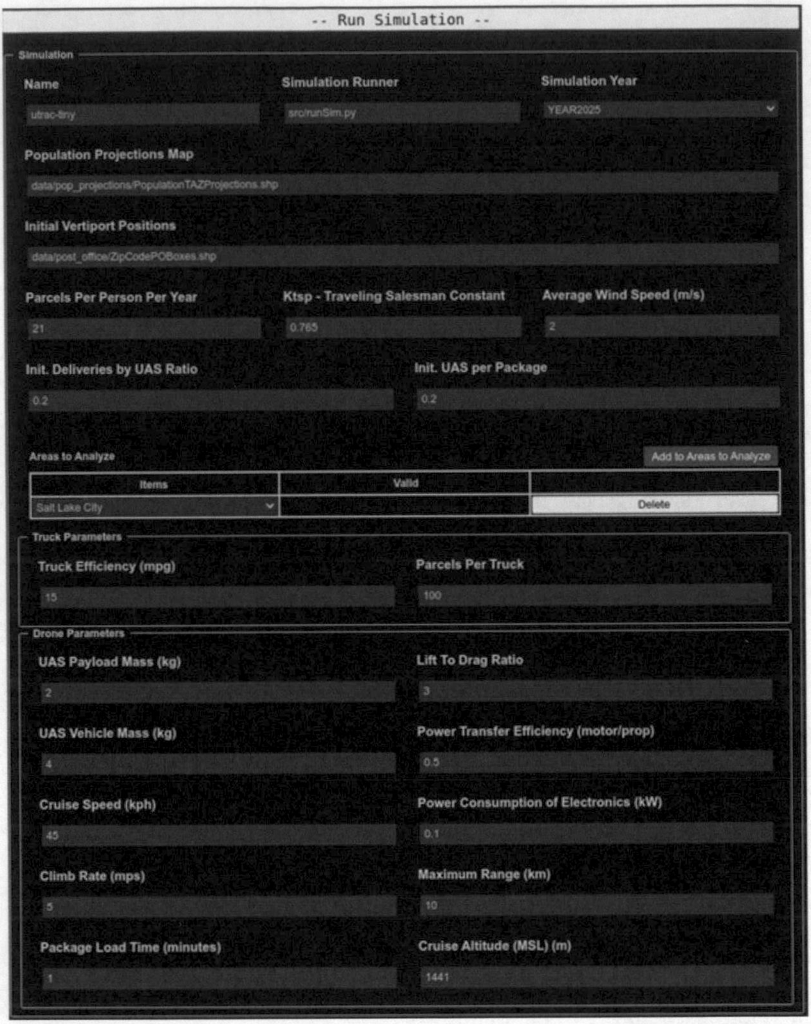

Fig. 9.7 The simulation form

9.3.3 Simulation Procedure

The following section describes each step of the simulation in detail. The complete source code of the simulation is delivered with UAAMS and accessible from the file explorer.

Load Input Location Data This step includes loading the following location data:

1. Vertiport Locations
2. Truck Depot Locations

3. Population Projections (at the TAZ level)
4. Building Locations

Depending on the size of the dataset, the location data can be loaded either from the filesystem or from a web-map server (one is provided with UAAMS). For example, the population projections dataset was small enough to have loaded from a shapefile located on the filesystem into memory, while the building locations dataset is too large to handle in this way. The buildings dataset is loaded into the web-map server offline, then during the simulation the server is queried within constrained areas.

Initialize UAS Model The UAS model considered for this simulation is derived from a study conducted by D'Andrea [29]. The parameters considered are listed in Table 9.1.

Filter TAZ Zones Based on Requested City Areas The simulation can consider a subset of areas in Utah, defined by the "City Areas" column in the Population Projections dataset [116].

Create Demand (Parcel Requests for a Day) For each TAZ under consideration, uniformly sample the building centroids within that TAZ from a binomial distribution with intensity given by the estimated number of parcels per person per day.

Source Parcel Requests to Nearest Vertiport For each parcel request in the demand dataset (from the last step), calculate the nearest vertiport that can serve that request.

Calculate Truck/UAS Job Mix Fixed by user input into the simulation form.

Generate UAS Trajectories For each parcel request that is served by a UAS, generate a trajectory based on the vehicle parameters provided in the simulation form.

Generate Animation To help visualize the distribution and density of drone flights that meet the demand specified in the simulation, a 3D animation (see figure below) is generated that can be viewed within the workspace. The animation is specified by an open-source human/machine readable file called CZML [6]. Figure 9.8 shows an example frame from this animation.

Generate Results Results are generated and stored in figures as described in Fig. 9.9.

9.4 Conclusions and Future Work

The implementation of a cloud-based collaborative simulator represents the bulk of this UTRAC project. A major unexpected hurdle was the need for significant memory and compute resources to execute the simulations. In particular, the

Fig. 9.8 One Frame of the simulation animation results

building dataset represented a challenge and required the use of a web-map server rather than simply loading the entire shapefile into memory. During the demand creation step in the simulation implementation, small areas were queried to avoid overwhelming memory resources.

With regard to the test simulation, the total energy required by both UAS and trucks combined was slightly more than the required energy if trucks were to meet the same demand alone. This could be a result of the chosen vehicle parameters, as well as the location distribution of vertiports. Despite this result, it may still be desirable to take advantage of the short distance and round-trip times that UAS has to offer.

Another challenge was encountered with the optimization objective. It was found to be computationally difficult to consider all the vertiport locations, as well as the various input parameters that could have been considered. The algorithm that was chosen to perform the optimization is called the Covariance-Matrix Adaptation Evolution Strategy [28] because it has a good reputation for handling highly non-linear problems with many variables. However, the algorithm proved to be rather slow, likely due to the random sampling and database querying that occur during the demand creation step. Consequently, in the time allotted for optimization, the algorithm did not successfully converge to a solution.

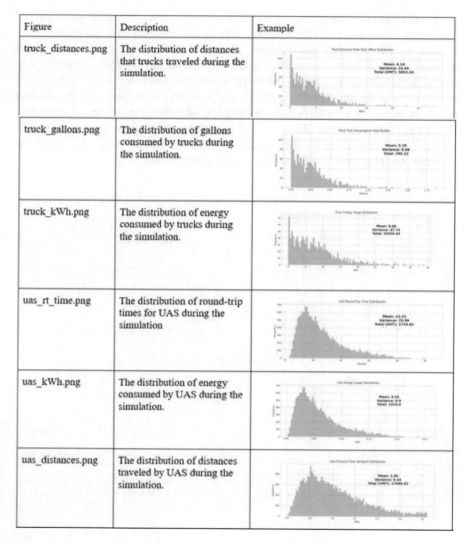

Figure	Description	Example
truck_distances.png	The distribution of distances that trucks traveled during the simulation.	
truck_gallons.png	The distribution of gallons consumed by trucks during the simulation.	
truck_kWh.png	The distribution of energy consumed by trucks during the simulation.	
uas_rt_time.png	The distribution of round-trip times for UAS during the simulation	
uas_kWh.png	The distribution of energy consumed by UAS during the simulation.	
uas_distances.png	The distribution of distances traveled by UAS during the simulation.	

Fig. 9.9 Simulation-produced plots

Overall, the implementation and deployment of the UAAMS was successful and will be available for continuing research in the development of advanced air mobility in the state of Utah. This type of tool is critical for research in this area, as it incorporates the latest software and infrastructure development techniques available. Also shown was the viability of considering the large-scale impacts (e.g., environmental) of advanced air mobility on specific communities by using micro-simulation technology. In contrast to micro-simulations performed for human-controlled ground and air traffic, in this case a model for the autonomous agents has the potential to be exact, since their algorithms must be documented.

Less certain are the emergent effects that may result from the parallel execution of these algorithms, and the environmental impacts of their collective behavior.

This UTRAC project has resulted in feedback from stakeholders that can be used to improve the simulation tool, as well as the resulting analysis. For example, simplifications to the workspace are necessary for non-developer, or non-engineer, stakeholders (e.g., planners). It is crucial to enable a frictionless communication between them and a tool like UAAMS has the potential to fill this role. Future improvements will also consider optimizations to the object function in the optimization procedure—this would allow a more thorough exploration of the possible configurations of vehicle and vertiport parameters.

Chapter 10
UAS Coalition Forces Coordination Scenario

10.1 Introduction

Consider now the problem of safely coördinating a set of multi-modal coalition forces' asset trajectories in a congested environment scenario. The lane-based UTM described in the preceding chapters can be extended to address this problem and will be called LEMANS-MM (Multi-Modal). Figure 10.1 illustrates the problem. As can be seen in the figure, this approach requires access to heterogeneous datasets that provide models of the various platforms involved, as well as the set of constraints and requirements imposed for tactical or strategic purposes, and a characterization of rogue or non-nominal trajectories. This allows for the determination of trajectories for high-speed projectiles, etc., which can move through the set of coalition agents without posing a threat (e.g., artillery rounds passing between aircraft).

The goals for the LEMANS-MM system are to:

- Quickly and automatically create lanes with desired topology.
- Use highly efficient algorithms to determine strategically deconflicted flight plans.
- Allow coalition forces to keep their flight data private when they deconflict.
- Provide strong support for handling contingencies (e.g., weather, platform failures, emergency use of airways, rogue platforms, etc.).
- Provide a visualization interface for lane operations to make it easy for operators to detect anomalies.
- Provide efficient and effective methods to pre-compute projectile trajectories through the lane system and deconflict them.
- Allow application of machine learning methods to determine best multi modal asset parameters (e.g., lane topology, lane speed limits, minimal headway distance, necessary lane volume, etc.).

© The Author(s), under exclusive license to Springer Nature Switzerland AG 2022   149
D. Sacharny, T. C. Henderson, *Lane-Based Unmanned Aircraft Systems Traffic Management*, Unmanned System Technologies,
https://doi.org/10.1007/978-3-030-98574-5_10

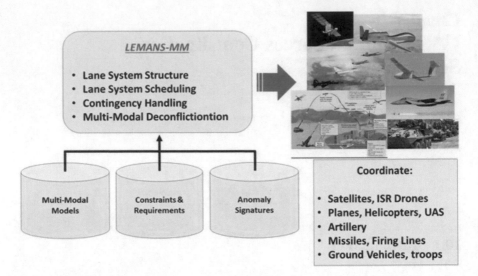

Fig. 10.1 The LEMANS-MM System will take information about the specific assets to be coördinated, including geographic, weather, asset priorities, and other constraints, as well as a set of known anomaly signatures

This approach requires the application of a multi-altitude lane set to allow the different types of aircraft to operate at appropriate altitudes. Figure 10.2 shows an example sketch with three separate lanes altitudes, in this case for Intelligence, Surveillance and Reconnaissance (ISR) platforms at the highest altitude, planes and helicopters at the mid-level, and UAS tactical drones at the lowest altitude. Given the GIS data, mission constraints and requirements, and rogue types, LEMANS-MM automatically generates the set of lanes for use by the manned and unmanned aircraft.

10.2   Airway Creation and Deconfliction

The coalition airways shown in Fig. 10.2 are just an example of what can be defined. In this particular case, the multiple levels are defined as three separate airways for the ISR, manned and unmanned UAS aircraft. Although this does impose the burden of strategic deconfliction in three separate lane systems, it has some advantages as well. First, if the three sets of launch and land lanes are defined so as not to interfere with each other (that is, they are well-separated), then there is little chance that nominal flights will have conflicting trajectories. Moreover, if one of the coalition forces wants to have its own private lane system, then it would just need to make sure that this new set of lanes does not conflict with the shared lane systems.

Multi-Domain Air Coordination Air Lane System

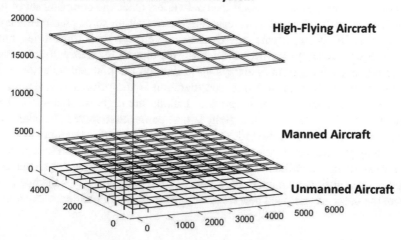

Fig. 10.2 Multi-altitude lane layout. Highest altitude is for ISR drones; mid-level altitude is for manned aircraft; lowest altitude is for UAS tactical drones

One reason to have separate lane systems for the different altitudes is that this allows each of them to be based off of different and more appropriate ground networks. For example, a specific ground network layout is not necessarily of much interest to the high-altitude surveillance flights, but for mid-level manned and low-level UAS aircraft there may be good reason to conform to (or avoid) local ground networks as well as to take into account local topography, including man-made structures such as buildings and towers. It is also likely that manned flights will require that certain areas be avoided (e.g., where there exists a high likelihood of anti-aircraft capability), whereas low-flying UAS may be tasked with tracking ground vehicles or other objects of interest.

On the other hand, if a single comprehensive set of lanes is desired, then it is possible to generate it, but it would also require the capability of restricting the usage of certain lanes to specific aircraft types. Moreover, if there is an allowance for shared lanes (across aircraft mission types), then air speed differences may pose problems since the speed of fighter planes is so much higher than, say, standard UAS drone platforms.

The next issue is the determination of flight paths through the air lane network. Contrary to the package delivery application, the different types of missions here require different considerations in order to be effective. For example, high-altitude reconnaissance missions may spend up to several hours acquiring data and delivering it through satellite links. Also, the number of ISR aircraft is minimal compared to the other types of missions. For manned aircraft the major motivation for airway lanes is to allow coördination while they are en route to or returning from the active engagement area, where of course, their movement would not be pre-planned by means of an existing set of lanes. Finally, low-altitude UAS platforms may be used to support other actions, e.g., by means of reconnaissance or monitoring, as well as for small load deliveries. Once the complete set of flights is defined, then the nominal trajectories are known within some predictable level of uncertainty. Of course, flights may dynamically change their trajectories, but this would require making a new reservation for the altered trajectory. In this way, the traffic management system is always up-to-date on the planned trajectories of all flights. Another thing to take into consideration is the priority of each coalition member, or the type of flight, in terms of making the flight reservation. Generally speaking the flight reservation system is first-come, first-served. In other words, lanes will be allocated in the order of the requests, and later attempts to schedule flights may require that they be delayed from their requested launch time in order to achieve strategic deconfliction. It is also possible to assign priorities so that certain flights (or coalition members) may preëmpt existing, already scheduled flights where the latter are assigned new times of passage through their lane sequence.

10.3 Multi-Modal Activities

In a warfare scenario, there will be a combined set of air, ground, and possibly other forces working together to achieve strategic and tactical goals. This may require disruption of the airways for other uses. Here we consider the case of artillery fire where the projectiles follow a parabolic course through the airway lanes in order to get to their targets. Most field artillery has a range of several kilometers and can reach similar altitudes as well. While it is possible to account for other types of trajectories, e.g., air-to-air missiles, and strategically deconflict them with coalition force flights, we do not explore that capability here.

Consider the trajectory of a projectile. Given an initial velocity of v_0 mps with θ being the angle of fire, then the velocity is given by

$$v_x = v_0 cos(\theta)$$

$$v_y = v_0 sin(\theta) - gt$$

These equations do not take air resistance into account (to get an estimate of the impact of air resistance, scale the numbers given below by 0.56). The position of the projectile at time t after launch is

$$x = v_0 cos(\theta)$$

$$y = v_0 t sin(\theta) - \frac{1}{2}gt^2$$

The total time the projectile is in the air before returning to its launch altitude is given by

$$t_{total} = \frac{2v_0 sin(\theta)}{g}$$

The time it takes for the projectile to reach its maximum altitude is

$$t_h = \frac{v_0 sin(\theta)}{g}$$

and the maximum height for a given initial velocity v_0 and launch angle θ is

$$h = \frac{v_0^2 sin^2(\theta)}{2g}$$

Finally, in order to reach location (x, y, z) from location $0, 0, 0$ with a projectile fired with initial velocity v_0, the required angle satisfies the equation:

$$tan(\theta) = \frac{v_0^2 \pm \sqrt{v_0^4 - g(g(x^2 + y^2) + 2zv_0^2)}}{g\sqrt{x^2 + y^2}}$$

It will be seen in the simulation experiment described below that a standard field artillery unit (an M777 howitzer in this case) will generally achieve the altitude of the highest airway lanes described previously; thus, it is an important capability to be able to deconflict artillery projectiles through the lane system.

10.4   Simulation Experiment

In order to demonstrate these methods, a scenario is presented that involves

- Three coalition force members (called Red, Green, and Blue).
- Bogey aircraft (called Black).
- Artillery projectiles (called Magenta).

The coalition forces share access to three distinct airway lane networks: ISR, manned, and UAS.

ISR Airway Lane Network The ISR lane network is a rectangular grid that covers an area of 5 km squared, has grid lines spaced at 1km intervals, has one launch and one land lane, and has its two-level lane set at 18,000 and 18,200 meters, respectively. There are 444 lanes and 252 lane vertexes. Note that the area from $x = 0$ to $x = 3000$ m is considered coalition-held territory while from $x = 3000$ to $x = 5000$ meters is enemy-held terrain.

Manned Airway Lane Network The manned air network is derived from rectangular grid ground network that covers the same 5 km squared area as the ISR network, has grid lines every 500 meters, has one launch and one land lane, and has its two-level lanes at 4000 meters and 4200 meters, respectively. There are 1574 lanes and 892 lane vertexes.

UAS Airway Lane Network The UAS air lane system is derived from a rectangular grid network that covers the same 5km squared area as the other two air networks, has grid spacings every 300 meters, has ten launch and ten land lanes, and has its two-level lanes at 600 meters and 620 meters, respectively. There are 1626 lanes and 944 lane vertexes.

There are five ISR flights per coalition member, and these are scheduled in the order Green, Red, and Blue. The flight path is the same for every flights; they are simply staggered in time. The mission is to ascend to 18,000 meters, make a pass around the outer perimeter of the ISR grid, then circle in the opposite direction, and then proceed to land.

The manned missions are planned similarly. Each coalition member schedules five flights. For this case, the mission proceeds from the launch site to a randomly selected interior vertex of the air lane network, and from there on to a vertex near the enemy area, then returning through another randomly selected interior vertex, and then on to the landing lane.

The UAS missions are similar to the manned flight missions. Each coalition member schedules five flights. The mission starts from a randomly selected launch site from among the ten possible, on to a randomly selected interior point, then on to a vertex near the enemy area, back through a randomly selected interior vertex, and finally on to a randomly selected landing lane from among the ten possible.

A set of five enemy aircraft (bogeys) are also introduced into the simulation; however, they are obviously not part of the scheduled flights, but rather their trajectories are generated as Rogue Type II flights (defined in Chap. 5). That is, they launch from a randomly selected ground location in enemy terrain, fly up to a randomly selected lane, and fly along it. This pattern is repeated a total of ten times: randomly choose a lane, fly to it, and then fly along it. After completing this, the flight lands at a randomly selected ground location in enemy territory. These flights are added just to demonstrate the capability to include non-scheduled flights in the simulation.

Finally, five hundred artillery rounds are planned:

- A random location is selected inside coalition-held terrain.
- A random location is selected in enemy terrain.
- The angle of fire is determined from the two positions.
- The trajectory of the projectile is determined.
- The trajectory is deconflicted by delaying the shot until no flight is endangered by the trajectory.

Angle of Fire Determination The angle is determined through the previously given equation, and then solving for θ using the *arctan* function. Note that in order to satisfy the conditions of the equation, it is necessary to translate the firing point,

X_1, and the target point X_2 by $-X_1$ so that the firing location of the howitzer is at the origin of the coördinate system.

Projectile Trajectory Determination The trajectory consists of a sequence of three-dimensional points and associated time of passage. The first element of the sequence is then the randomly selected firing location with the initial time as the desired firing time. A time step is selected for the calculation (in the simulation described below, a time step of 0.1 seconds is used). The trajectory consists of a sequence of locations and times:

$$\mathcal{T} = \{(X_1, t_1), (X_2, t_2), \ldots, (X_n, t_n)\}$$

Projectile Trajectory Deconfliction In order to deconflict the projectile trajectory, it is necessary to ensure that at no point in the trajectory does the projectile get too close to a flight in a lane passing at the same time. This is achieved using the following information:

- The ISR, manned and UAS lane network data.
- The ISR, manned and UAS kd-tree spatial database models (described in Chap. 5).
- The ISR, manned and UAS flight reservation information (i.e., the flight schedules).
- The minimum distance to maintain (i.e., the headway).
- The projectile trajectory, $\mathcal{T}$.

Given a trajectory, $\mathcal{T} = \{(X_1, t_1), (X_2, t_2), \ldots, (X_n, t_n)\}$, then to guarantee that it is safe, it is necessary to:

1. For every element of the sequence, (X_k, t_k), find all lanes in the three air networks that are within headway distance of the trajectory points (this is done using the spatial databases to find nearest points in the model).
2. For each such lane, determine if there is a flight that passes at such a time as to be unsafe (this is done by using the flight reservations to determine which flights are in the lane when the projectile passes, and then using the airways lane data to determine exactly where the flight is in the lane when the projectile passes).
3. If any such flight exists add some fixed delay time to the artillery shot firing time, and re-start the analysis of the trajectory.

10.5   Experimental Results

Given the three networks described above, the M777 Howitzer is used as an example artillery platform for the simulation. The M777 is produced by BAE systems, and has the following characteristics:

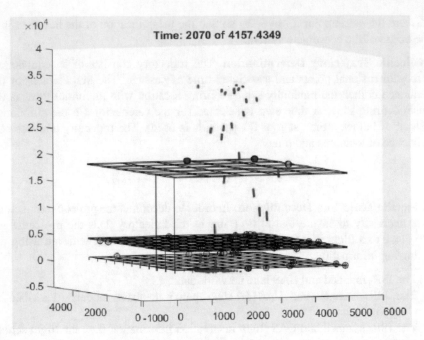

Fig. 10.3 One time step during the simulation of the artillery firing through the air lane networks where the artillery projectiles trajectories are deconflicted with the scheduled coalition flights

- Weighs about 4000 kg.
- Has an effective range of 24 km.
- Has a muzzle velocity of 827 m/s.
- Has a maximum angle of fire of 71.7°.

Thus, the maximum projectile altitude is about 31,000 meters when air resistance is not considered, and about 18,000 meters when it is. In the simulation, air resistance is not taken into account, and there is no constraint on firing angle.

Figure 10.3 shows the simulation at time step $t = 2070$. The magenta projectile tracks have all been deconflicted, and their paths through the lane systems pose no safety threats for the coalition forces flights. The Green, Red, and Blue forces flights can be seen with corresponding asterisk colors inside a black circle. The bogey flights are indicated as black asterisks inside a clack circle. Five hundred rounds of artillery projectiles go through the air networks during the simulation time.

Areas for exploration in future work on this topic include:

- Creation of a single combined airway network in which lane usage may be restricted by flight type or coalition member.
- The assignment of priorities to coalition members.
- The simulation and analysis of air-to-air trajectories.

- A more full-blown simulation of mission sets for the various forces; e.g., coördinated reconnaissance and combat action, response to bogeys, using UAS to track ground vehicles, etc.

The incorporation of these capabilities into a physics-based game engine would allow the evaluation of strategic and tactical plans in realistic scenarios.

Appendix A
Space–Time Lane Diagram Enumeration

See Figs. A.1, A.2, A.3, A.4, A.5, A.6, A.7, A.8, A.9, and A.10.

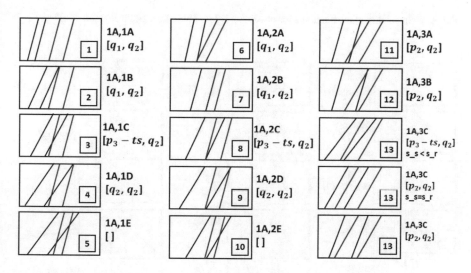

Fig. A.1 Space–time lane diagrams for possible label combinations 1 through 13

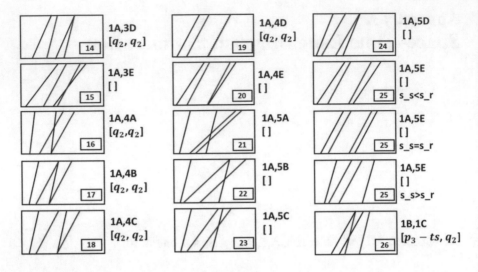

Fig. A.2 Space–time lane diagrams for possible label combinations 14 through 26

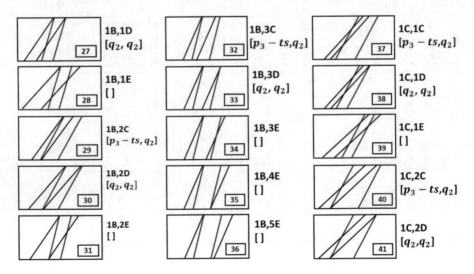

Fig. A.3 Space–time lane diagrams for possible label combinations 27 through 41

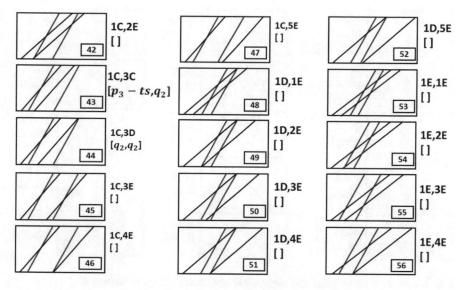

Fig. A.4 Space–time lane diagrams for possible label combinations 42 through 56

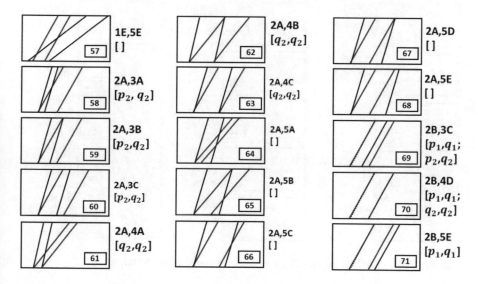

Fig. A.5 Space–time lane diagrams for possible label combinations 57 through 71

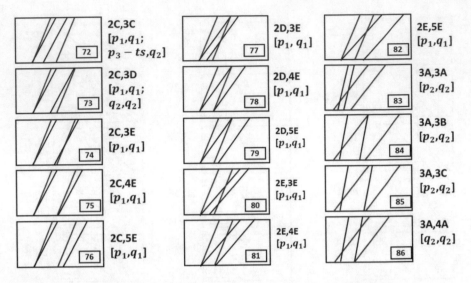

Fig. A.6 Space–time lane diagrams for possible label combinations 72 through 86

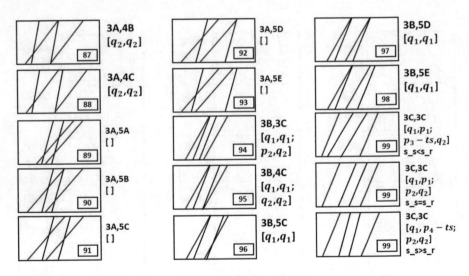

Fig. A.7 Space–time lane diagrams for possible label combinations 87 through 99

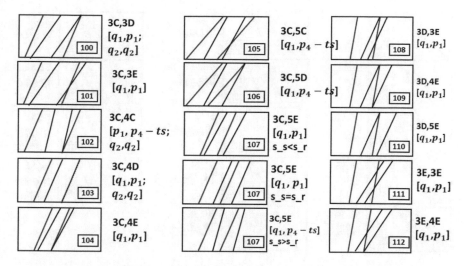

Fig. A.8 Space–time lane diagrams for possible label combinations 100 through 112

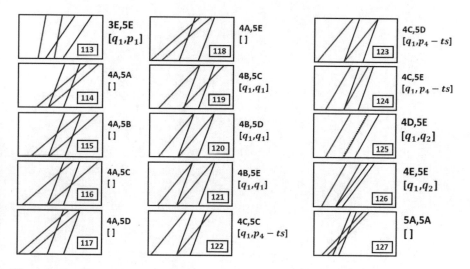

Fig. A.9 Space–time lane diagrams for possible label combinations 113 through 127

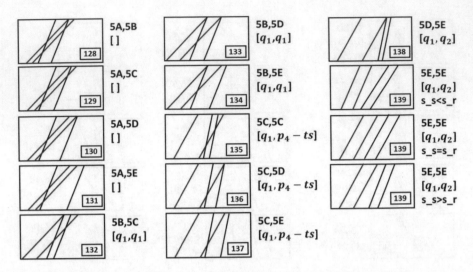

Fig. A.10 Space–time lane diagrams for possible label combinations 128 through 139

Appendix B
Matlab Code for Algorithm LBSD

```
function possible = UR_possible_times_int(possible0,speed,cor_
    list,...cor_lengths,flights,ht)
% UR_possible_times_int - provide possible strategically
    deconflicted
%        launch time intervals given a requested interval and the
%        scheduled flights
% On input:
%     possible0 (1x2 vector): first and last possible launch
        times speed (float): speed to requesting UAS
%     cor_list (kx1 vector): list of corridors to be traversed
        (in order)
%     cor_lengths (kx1 vector): lengths of corridors to be
        traversed
%     flights (vector struct): scheduled flights (given per
        corridor)
%     ht (float): headway time
% On output:
%     possible (nx2 array): each row is a continuous interval of
%     possible
%        starting flight times
% Call:
%     inters =
%     UR_possible_times_int([4,35],2,[13,6,14],[500,6,500],fl,5);
% Author:
%     T. Henderson
%     UU
%     Summer 2019
%

len_cor_list = length(cor_list);
intervals = possible0;
offset = 0;
c = 0;
total_time = 0;
```

© The Author(s), under exclusive license to Springer Nature Switzerland AG 2022   165
D. Sacharny, T. C. Henderson, *Lane-Based Unmanned Aircraft Systems Traffic
Management*, Unmanned System Technologies,
https://doi.org/10.1007/978-3-030-98574-5

```matlab
while ~isempty(intervals)&c<len_cor_list
    c = c + 1;
    dc = cor_lengths(c);
    cor = cor_list(c);
    ts = dc/speed;
    [num_intervals,dummy] = size(intervals);
    for k = 1:num_intervals
        intervals(k,:) = intervals(k,:) + [offset,offset];
    end
    c_flights = flights(cor).flights;
    if ~isempty(c_flights)
        [num_c_flights,dummy] = size(c_flights);
        f = 0;
        [num_intervals,dummy] = size(intervals);
        while f<num_c_flights&~isempty(intervals)
            f = f + 1;
            tr1 = min(intervals(:,1));
            tr2 = max(intervals(:,2));
            ts1 = c_flights(f,1);
            ts2 = c_flights(f,2);
            tr1e = tr1 + dc/speed;
            tr2e = tr2 + dc/speed;
            if ~((ts1+ht<=tr1&ts2+ht<=tr1e)|(ts1-ht>=tr2&ts2-ht>
              =tr2e))
                new_intervals = [];
                for k = 1:num_intervals
                    k_intervals = UR_OK_sched_req_enum(c_flights
                    (f,1),...
                        c_flights(f,2),c_flights(f,3),intervals
                        (k,1),...
                        intervals(k,2),speed,dc,ht);
                    new_intervals = UR_merge_intervals(k_inter-
                    vals,...
                        new_intervals);
                end
                intervals = new_intervals;
                if isempty(intervals)
                    num_intervals = 0;
                else
                    num_intervals = length(intervals(:,1));
                end
            end
        end
    end
    offset = ts;
    total_time = total_time + ts;
end
total_time = total_time - ts;
[num_intervals,dummy] = size(intervals);
for k = 1:num_intervals
    intervals(k,:) = intervals(k,:) - [total_time,total_time];
end
if ~isempty(intervals)
    t1 = intervals(1,1);
```

```
        offset = 0;
        for c = 1:len_cor_list
            cor = cor_list(c);
            if ~isempty(flights(cor).flights)...
                    &abs(t1-flights(cor).flights(1,1))<7
                tch = 0;
            end
            t1 = t1 + cor_lengths(c)/speed;
        end
end

%   return all intervals
possible = intervals;
return

function intervals =
UR_OK_sched_req_enum(ts1,ts2,s_s,tr1,tr2,s_r,d,ht)
% UR_OK_sched_req_enum - determine OK intervals for proposed
                        flight in
%                       specific corridor
% On input:
%     ts1 (float): start of scheduled flight
%     ts2 (float): end of scheduled flight
%     s_s (float): speed of scheduled flight
%     tr1 (float): min start time requested
%     tr2 (float): max start time requested
%     s_r (float): speed of requested flight
%     d (float): corridor length
%     ht (float): headway time
% On output:
%     intervals (nx2 array): possible start time intervals
% Call:
%     int1 = UR_OK_sched_req_enum(23,51,5,8,40,3,49,5);
% Author:
%     T. Henderson
%     UU
%     Summer 2019
%

persistent first itable

if isempty(first)
    first = 0;
    itable = [...
        1 1 1 1;...  % Case   1
        1 1 1 2;...  % Case   2
        1 1 1 3;...  % Case   3
        1 1 1 4;...  % Case   4
        1 1 1 5;...  % Case   5
        1 1 2 1;...  % Case   6
        1 1 2 2;...  % Case   7
        1 1 2 3;...  % Case   8
        1 1 2 4;...  % Case   9
        1 1 2 5;...  % Case  10
```

```
1 1 3 1;...   % Case   11
1 1 3 2;...   % Case   12
1 1 3 3;...   % Case   13
1 1 3 4;...   % Case   14
1 1 3 5;...   % Case   15
1 1 4 1;...   % Case   16
1 1 4 2;...   % Case   17
1 1 4 3;...   % Case   18
1 1 4 4;...   % Case   19
1 1 4 5;...   % Case   20
1 1 5 1;...   % Case   21
1 1 5 2;...   % Case   22
1 1 5 3;...   % Case   23
1 1 5 4;...   % Case   24
1 1 5 5;...   % Case   25
1 2 1 3;...   % Case   26
1 2 1 4;...   % Case   27
1 2 1 5;...   % Case   28
1 2 2 3;...   % Case   29
1 2 2 4;...   % Case   30
1 2 2 5;...   % Case   31
1 2 3 3;...   % Case   32
1 2 3 4;...   % Case   33
1 2 3 5;...   % Case   34
1 2 4 5;...   % Case   35
1 2 5 5;...   % Case   36
1 3 1 3;...   % Case   37
1 3 1 4;...   % Case   38
1 3 1 5;...   % Case   39
1 3 2 3;...   % Case   40
1 3 2 4;...   % Case   41
1 3 2 5;...   % Case   42
1 3 3 3;...   % Case   43
1 3 3 4;...   % Case   44
1 3 3 5;...   % Case   45
1 3 4 5;...   % Case   46
1 3 5 5;...   % Case   47
1 4 1 5;...   % Case   48
1 4 2 5;...   % Case   49
1 4 3 5;...   % Case   50
1 4 4 5;...   % Case   51
1 4 5 5;...   % Case   52
1 5 1 5;...   % Case   53
1 5 2 5;...   % Case   54
1 5 3 5;...   % Case   55
1 5 4 5;...   % Case   56
1 5 5 5;...   % Case   57
2 1 3 1;...   % Case   58
2 1 3 2;...   % Case   59
2 1 3 3;...   % Case   60
2 1 4 1;...   % Case   61
2 1 4 2;...   % Case   62
2 1 4 3;...   % Case   63
2 1 5 1;...   % Case   64
```

```
2  1  5  2;...    % Case   65
2  1  5  3;...    % Case   66
2  1  5  4;...    % Case   67
2  1  5  5;...    % Case   68
2  2  3  3;...    % Case   69
2  2  4  4;...    % Case   70
2  2  5  5;...    % Case   71
2  3  3  3;...    % Case   72
2  3  3  4;...    % Case   73
2  3  3  5;...    % Case   74
2  3  4  5;...    % Case   75
2  3  5  5;...    % Case   76
2  4  3  5;...    % Case   77
2  4  4  5;...    % Case   78
2  4  5  5;...    % Case   79
2  5  3  5;...    % Case   80
2  5  4  5;...    % Case   81
2  5  5  5;...    % Case   82
3  1  3  1;...    % Case   83
3  1  3  2;...    % Case   84
3  1  3  3;...    % Case   85
3  1  4  1;...    % Case   86
3  1  4  2;...    % Case   87
3  1  4  3;...    % Case   88
3  1  5  1;...    % Case   89
3  1  5  2;...    % Case   90
3  1  5  3;...    % Case   91
3  1  5  4;...    % Case   92
3  1  5  5;...    % Case   93
3  2  3  3;...    % Case   94
3  2  4  3;...    % Case   95
3  2  5  3;...    % Case   96
3  2  5  4;...    % Case   97
3  2  5  5;...    % Case   98
3  3  3  3;...    % Case   99
3  3  3  4;...    % Case  100
3  3  3  5;...    % Case  101
3  3  4  3;...    % Case  102
3  3  4  4;...    % Case  103
3  3  4  5;...    % Case  104
3  3  5  3;...    % Case  105
3  3  5  4;...    % Case  106
3  3  5  5;...    % Case  107
3  4  3  5;...    % Case  108
3  4  4  5;...    % Case  109
3  4  5  5;...    % Case  110
3  5  3  5;...    % Case  111
3  5  4  5;...    % Case  112
3  5  5  5;...    % Case  113
4  1  5  1;...    % Case  114
4  1  5  2;...    % Case  115
4  1  5  3;...    % Case  116
4  1  5  4;...    % Case  117
4  1  5  5;...    % Case  118
```

```
                  4  2  5  3;...    % Case 119
                  4  2  5  4;...    % Case 120
                  4  2  5  5;...    % Case 121
                  4  3  5  3;...    % Case 122
                  4  3  5  4;...    % Case 123
                  4  3  5  5;...    % Case 124
                  4  4  5  5;...    % Case 125
                  4  5  5  5;...    % Case 126
                  5  1  5  1;...    % Case 127
                  5  1  5  2;...    % Case 128
                  5  1  5  3;...    % Case 129
                  5  1  5  4;...    % Case 130
                  5  1  5  5;...    % Case 131
                  5  2  5  3;...    % Case 132
                  5  2  5  4;...    % Case 133
                  5  2  5  5;...    % Case 134
                  5  3  5  3;...    % Case 135
                  5  3  5  4;...    % Case 136
                  5  3  5  5;...    % Case 137
                  5  4  5  5;...    % Case 138
                  5  5  5  5];      % Case 139
end

intervals = [];

t_across = d/s_r;
%t_across = ceil(d/s_r);
p1 = ts1 - ht;
p2 = ts1 + ht;
p3 = ts2 + ht;
p4 = ts2 - ht;
q1 = tr1;
q2 = tr2;
q3 = tr2 + t_across;
q4 = tr1 + t_across;

if p1<q1
    i1 = 1;
elseif p1==q1
    i1 = 2;
elseif p1>q1&p1<q2
    i1 = 3;
elseif p1==q2
    i1 = 4;
else
    i1 = 5;
end
if p2<q1
    i3 = 1;
elseif p2==q1
    i3 = 2;
elseif p2>q1&p2<q2
    i3 = 3;
elseif p2==q2
```

```
        i3 = 4;
else
        i3 = 5;
end

if p3<q4
    i4 = 1;
elseif p3==q4
    i4 = 2;
elseif p3>q4&p3<q3
    i4 = 3;
elseif p3==q3
    i4 = 4;
else
    i4 = 5;
end
if p4<q4
    i2 = 1;
elseif p4==q4
    i2 = 2;
elseif p4>q4&p4<q3
    i2 = 3;
elseif p4==q3
    i2 = 4;
else
    i2 = 5;
end

index = find(itable(:,1)==i1&itable(:,2)==i2&itable(:,3)==i3...
    &itable(:,4)==i4);
switch index
    case {1,2,6,7,125,126,138,139}
        intervals = [q1,q2];
    case {3,8,26,29,32,37,40,43}
        intervals = [p3-t_across,q2];
    case {4,9,14,16,17,18,19,27,30,33,38,41,44,61,62,63,86,87,88}
        intervals = [q2,q2];
    case {11,12,58,59,60,83,84,85}
        intervals = [p2,q2];
    case 13
        if s_s<s_r
            intervals = [p3-t_across,q2];
        else
            intervals = [p2,q2];
        end
    case 69
        intervals = [p1,q1; p2,q2];
    case {70,73}
        intervals = [p1,q1; q2,q2];
    case {71,74,75,76,77,78,79,80,81,82}
        intervals = [p1,q1];
    case 72
        intervals = [p1,q1; p3-t_across,q2];
    case {94,95}
```

```
            intervals = [q1,q1;q2,q2];
    case 99
        if s_s<s_r
            intervals = [q1,p1; p3-t_across,q2];
        elseif s_s==s_r
            intervals = [q1,p1; p2,q2];
        else
            intervals = [q1,p4-t_across; p2,q2];
        end
    case {96,97,98,119,120,121,132,133,134}
        intervals = [q1,q1];
    case {100,103}
        intervals = [q1,p1; q2,q2];
    case 102
        intervals = [p1,p4-t_across; q2,q2];
    case {101,104,108,109,110,111,112,113}
        intervals = [q1,p1];
    case 107
        if s_s<=s_r
            intervals = [q1,p1];
        else
            intervals = [q1,p4-t_across];
        end
    case {105,106,122,123,124,135,136,137}
        intervals = [q1,p4-t_across];
end

function new_int = UR_merge_intervals(int1,int2)
% UR_merge_intervals - given two sets of intervals, merge them
% On input:
%     int1 (n1x2 array): first set of intervals
%     int2 (n2x2 array): second set of intervals
% On output:
%     new_int (px2 array): intersection of two interval sets
% Call:
%     new_int = UR_merge_intervals (int1,int2);
% Author:
%     T. Henderson
%     UU
%     Summer 2019
%

if isempty(int1)&isempty(int2)
    new_int = [];
    return
end

new_int = [int1;int2];
[vals,indexes] = sort(new_int(:,1));
new_int = new_int(indexes,:);
change = 1;
while change==1
    change = 0;
    len_new_int = length(new_int(:,1));
```

```
    for k = 1:len_new_int-1
        if new_int(k,1)==new_int(k+1,1)
            v_min = min(new_int(k,2),new_int(k+1,2));
            new_int(k+1,2) = v_min;
            new_int(k,:) = [];
            change = 1;
            break
        end
    end
end
```

Appendix C
Sample ABMS LBSD Code

The ABMS code uses MATLAB's object oriented interface. The following demonstrates some of the LBSD interfaces and the central data structures required by the lane-based approach.

Lane System

```
// The following code shows how to initialize a
// simple roundabout and plot the result
lane_length_m = 10;
altitude_m = 15;
lbsd = LBSD.genSampleLanes(lane_length_m,altitude_m);
plot(lbsd);
```

© The Author(s), under exclusive license to Springer Nature Switzerland AG 2022
D. Sacharny, T. C. Henderson, *Lane-Based Unmanned Aircraft Systems Traffic Management*, Unmanned System Technologies,
https://doi.org/10.1007/978-3-030-98574-5

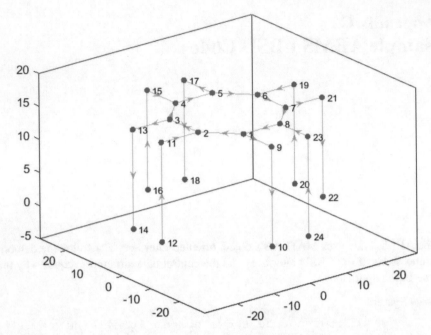

Lane IDs

Lane identifiers are associated with edges in the directed graph that represents the lane network. It is common practice in MATLAB to use integer indexes to increase performance; however, strings were selected for lane identifiers. This selection was made to enable the flexibility of delcting or adding lanes without having to reorder existing lanes.

```
lane_ids = lbsd.getLaneIds
```

```
lane_ids = 24x1 string
"2"
"9"
"3"
"4"
"13"
"5"
"6"
"17"
"7"
"8"
```

Lane Node Table
A MATLAB table holds data regarding the nodes in the directed graph that
represents the lane system. The critical fields include the 3-dimensional position
of the node, and whether it is a launch or land location.

```
lbsd.lane_graph.Nodes
```

	XData	YData	ZData	Launch	Land	Name
1	-5	-12.0711	15	0	0	'1'
2	-12.0711	-5	15	0	0	'2'
3	-12.0711	5	15	0	0	'3'
4	-5	12.0711	15	0	0	'4'
5	5	12.0711	15	0	0	'5'
6	12.0711	5	15	0	0	'6'
7	12.0711	-5	15	0	0	'7'
8	5	-12.0711	15	0	0	'8'
9	-5	-22.0711	15	0	0	'9'
10	-5	-22.0711	0	0	1	'10'
11	-22.0711	-5	15	0	0	'11'
12	-22.0711	-5	0	1	0	'12'
13	-22.0711	5	15	0	0	'13'
14	-22.0711	5	0	0	1	'14'

Lane Edge Table
The edge table represents the lanes in the directed graph and lists the names of
the nodes that are connected in the order that they may be traversed. A weight is
assigned to each edge, and in this case it is the Euclidean distance covered by the
edge.

```
lbsd.lane_graph.Edges
```

Random Reservation Generation
For demonstration and testing purposes, random lane reservations may be scheduled.

	EndNodes		Weight
1	'1'	'2'	10
2	'1'	'9'	10
3	'2'	'3'	10
4	'3'	'4'	10
5	'3'	'13'	10
6	'4'	'5'	10
7	'5'	'6'	10
8	'5'	'17'	10
9	'6'	'7'	10
10	'7'	'8'	10
11	'7'	'21'	10
12	'8'	'1'	10
13	'9'	'10'	15
14	'11'	'2'	10

```
start_time = 0;
end_time = 100;
lane_ids = ["1","2","3"];
num_res = 50;
speed = 1;
headway = 5;
lbsd.genRandReservations(start_time, end_time, num_
     res, ...lane_ids, speed, headway);
```

Reservations Table

The reservations table contains all the reservations that have been made on the lane
system. The critical fields include the entry time, exit time, speed, and headway (hd).
Since the exit time can be derived from the speed and entry time, it is not necessary
but is included for optimization reasons. Other underlying optimizations include
lane reservation lookup tables; MATLAB tables have performance drawbacks
when certain indexing operations are performed, for example lookup by the string
lane identifier. To combat these drawbacks and still enable the clean application
programming interface provided by the table data structure, several redundant
structures provide fast lookup and insertion for reservations.

```
lbsd.getReservations
```

	id	lane_id	entry_time_s	exit_time_s	speed	hd
1	"1"	"1"	98.7935	108.7935	1	5
2	"2"	"1"	17.0432	27.0432	1	5
3	"3"	"1"	25.7792	35.7792	1	5
4	"4"	"1"	39.6799	49.6799	1	5
5	"5"	"1"	7.3995	17.3995	1	5
6	"6"	"1"	68.4096	78.4096	1	5
7	"7"	"1"	62.0672	72.0672	1	5
8	"8"	"1"	75.8112	85.8112	1	5
9	"9"	"1"	87.1111	97.1111	1	5
10	"10"	"1"	53.0629	63.0629	1	5
11	"11"	"1"	33.5311	43.5311	1	5
12	"12"	"1"	45.2593	55.2593	1	5
13	"13"	"1"	93.5731	103.5731	1	5
14	"14"	"2"	98.7935	108.7935	1	5

Lane Reservation Lookup

An object of the LBSD class enables users to lookup the reservations for any lane.

```
lbsd.getLaneReservations("1")
```

	id	lane_id	entry_time_s	exit_time_s	speed	hd
1	"1"	"1"	98.7935	108.7935	1	5
2	"2"	"1"	17.0432	27.0432	1	5
3	"3"	"1"	25.7792	35.7792	1	5
4	"4"	"1"	39.6799	49.6799	1	5
5	"5"	"1"	7.3995	17.3995	1	5
6	"6"	"1"	68.4096	78.4096	1	5
7	"7"	"1"	62.0672	72.0672	1	5
8	"8"	"1"	75.8112	85.8112	1	5
9	"9"	"1"	87.1111	97.1111	1	5
10	"10"	"1"	53.0629	63.0629	1	5
11	"11"	"1"	33.5311	43.5311	1	5
12	"12"	"1"	45.2593	55.2593	1	5
13	"13"	"1"	93.5731	103.5731	1	5

Reservation Clearing

The LBSD class allows for resetting the reservations table.

```
lbsd.clearReservations()
```

Event Triggering

An event system is provided by the LBSD class to enable external services to react to different method calls. For example, this feature is used by the Air Traffic Operations Center (ATOC) class to update different plots whenever a UAS agent schedules a flight. Below is an example of how this feature can be used to print to the MATLAB terminal anytime a reservation is made.

Subscribe to a Reservation Events

```
lbsd.subscribeToNewReservation(\
     @(src,evt)disp("Reservation Made!"));
```

Since the random reservations method sequentially schedules flights, this method can be used to trigger the reservation events for testing, as in the example below.

```
start_time = 0;
end_time = 100;
lane_ids = ["1","2","3"];
num_res = 50;
speed = 1;
headway = 5;
lbsd.genRandReservations(start_time, end_time, num_
     res, \lane_ids, speed, headway);
```

```
Reservation Made!
Reservation Made!
Reservation Made!
...
```

Space–Time Lane Diagram

Both the LBSD class and the ATOC class have the ability to create space–time lane diagrams (STLD) for visualizing reservations. The example below shows how the ATOC object is constructed with a reference (a MATLAB handle) to an LBSD object, then constructs the STLD for several lanes by using the LBSD application programming interface.

```
atoc = ATOC(lbsd, 100, zeros(3), 90);
atoc.laneGraphs(lane_ids, [start_time, end_time]);
```

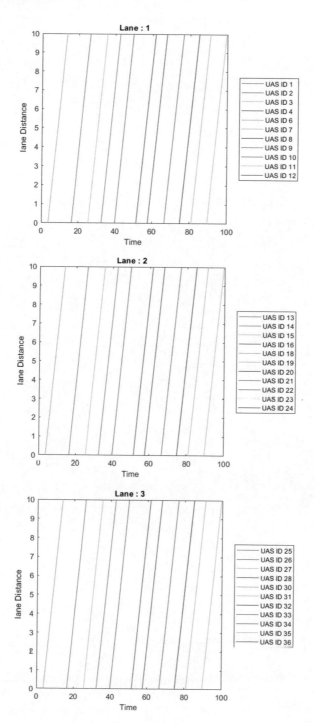

Appendix D
Abbreviations

AAM	Advanced Air Mobility
ABMS	Agent-Based Modeling and Simulation
ADS-B	Automatic Dependent Surveillance-Broadcast
AFOSR	Air Force Office of Scientific Research
AGL	Above Ground Level
AGRC	Automated Geographic Reference Center
ANSP	Air Navigation Service Provider
ARMD	Aeronautics Research Mission Directorate
ATC	Air Traffic Control
ATOC	Air Traffic Operations Center
BC	Betweenness Centrality
BDI	Belief, Desire, Intention
BSP	Bertsimas and Stock-Patternson
CMAES	Covariance Matrix Adaptation Evolution Strategy
CNF	Conjunctive Normal Form
CONOPS	Concept of Operations
CPAD	Closest Point of Approach Deconfliction
DDDAS	Data-Driven Dynamic Applications Systems
DSRC	Dedicated Short-Range Communications
DSS	Discovery and Synchronization of Services
FAA	Federal Aeronautics Administration
FIMS	Flight Information Management System
FNSD	FAA-NASA Strategic Deconfliction
GIS	Geographic Information System
GTMS	Ground Transportation Management System
ICAO	International Civil Aviation Organization
IGNRN	Integrated Graph of Natural Road Networks
ISR	Intelligence, Surveillance, and Reconnaissance
LBSD	Lane-Based Strategic Deconfliction

© The Author(s), under exclusive license to Springer Nature Switzerland AG 2022
D. Sacharny, T. C. Henderson, *Lane-Based Unmanned Aircraft Systems Traffic
Management*, Unmanned System Technologies,
https://doi.org/10.1007/978-3-030-98574-5

NAB	Nominal vs. Anomalous Behavior
NASA	National Aeronautics and Space Administration
NP	Non-deterministic Polynomial Time Complexity
NRE	Non-Recurring Engineering
MDP	Markov Decision Process
PSSAT	Probabilistic Sentence Satisfiability
PSU	Providers of Services for Urban Air Mobility
RL	Reinforcement Learning
RRT	Rapidly Exploring Random Trees
RTK	Real-Time Kinetic
SAT	Satisfiability
SBDC	Small Business Development Center
STLD	Space–Time Lane Diagram
sUAS	small Unmanned Aircraft Systems
TAP	Traffic Assignment Problem
TAZ	Traffic Analysis Zone
TCAS	Traffic Alert and Collision Avoidance System
TCL	Technical Capability Level
TFMP	Traffic Flow Management Problem
TFMRP	Traffic Flow Management and Rerouting Problem
TFR	Temporary Flight Restriction
TOA	Time of Arrival
TOD	Time of Departure
UAAMS	Utah Advanced Air Mobility System
UAM	Urban Air Mobility
UAS	Unmanned Aircraft Systems
UAV	Unmanned Aircraft Vehicles
UDOT	Utah Department of Transportation
UGRC	Utah Geospatial Resource Center
UQ	Uncertainty Quantization
USS	UAS Service Supplier
UTM	UAS Traffic Management
VM	Virtual Machine
VOR	Very High Frequency Omni-directional Range

Bibliography

1. F. Ahmadzai, K.M.L. Rao, S. Ulfat, Assessment and modeling of urban road networks using integrated graph of natural road networks (a GIS-based approach). J. Urban Manage. **8**, 109–125 (2019)
2. AirMap Company, Five critical enablers for safe, efficient, and viable UAS traffic management (UTM), in Whitepaper, Santa Monica (2018)
3. D. Alejo, J.M. Díaz-Báñez, J.A. Cobano, P. Pérez-Lantero, A. Ollero, The velocity assignment problem for conflict resolution with multiple aerial vehicles sharing airspace. J. Intell. Robot. Syst. Theory Appl. **69**(1–4), 331–346 (2013)
4. T. Alsinet, C.I. Chesnevar, L. Godo, G.R. Simari, A logic programming framework for possibilistic argumentation. Fuzzy Sets Syst. **159**(10), 1208–1228 (2008)
5. R. Alur, Formal verification of hybrid systems, in *Proceedings of the Ninth ACM International Conference on Embedded Software (EMSOFT)* (ACM Press, New York, 2011), pp. 273–278
6. CZML Guide AnalyticalGraphicsInc/czml-writer Wiki. (n.d.). GitHub. Retrieved June 1, 2019, from https://github.com/AnalyticalGraphicsInc/czml-writer/wiki/CZML-Guide
7. A. Arisha, P. Young, M. El Baradie, Job shop scheduling problem: an overview, in *International Conference for Flexible Automation and Intelligent Manufacturing (FAIM'01), Dublin* (2001), pp. 682–693
8. GitHub - astm-utm/Protocol: ASTM UTM Protocol (API and sequence diagrams). (n.d.). GitHub. Retrieved June 1, 2020., from https://github.com/astm-utm/Protocol
9. P.-L. Bacon, J. Harb, D. Precup, The option-critic architecture, in *The Thirty-First AAAI Conference on Artificial Intelligence, San Francisco* (2017)
10. J. Baculi, C. Ippolito, Onboard decision-making for nominal and contingency sUAS flight, in *AIAA Scitech 2019 Forum* (American Institute of Aeronautics and Astronautics, Reston, 2019)
11. S. Balachandran, C. Munoz, M.C. Consiglio, Implicitly coordinated detect and avoid capability for safe autonomous operation of small UAS, in *17th AIAA Aviation Technology, Integration, and Operations Conference* (American Institute of Aeronautics and Astronautics, Reston, 2017)
12. J. Barbagello, Instrument procedures handbook. Technical Report FAA-H-8083-16B, Federal Aviation Administration, Washington (2017)
13. M. Barthelemy, *Morphogenesis on Spatial Networks* (Springer, Berlin, 2018)
14. D. Bertsimas, S.S. Patterson, The air traffic flow management problem with enroute capacities. Oper. Res. **46**(3), 406–422 (1998)

© The Author(s), under exclusive license to Springer Nature Switzerland AG 2022
D. Sacharny, T. C. Henderson, *Lane-Based Unmanned Aircraft Systems Traffic Management*, Unmanned System Technologies,
https://doi.org/10.1007/978-3-030-98574-5

15. M. Biba, Integrating logic and probability: algorithmic improvements in Markov logic networks. Ph.D. Thesis, University of Bari, Bari (2009)
16. H.A.P. Blom, G.J. Bakker, Conflict probability and incrossing probability in air traffic management, in *Proceedings of the 41st IEEE Conference on Decision and Control, 2002, Las Vegas*, (2002), pp. 2421–2426
17. B. Boehm, C. Abts, S. Chulani, Software development cost estimation approaches – a survey. Ann. Softw. Eng. **10**(1), 177–205 (2000)
18. G. Boole, Further observations on the theory of probabilities. Lond. Edinburgh Dublin Philos. Mag. J. Sci. **2**, 96–101 (1851)
19. G. Boole, *An Investigation of the Laws of Thought* (Walton and Maberly, London, 1857)
20. Booz Allen Hamilton Inc., Kimley-Horn and Associates Inc., Transportation management center business planning and plans handbook. Technical Report TMC Pooled Fund Study, Federal Highway Administration, Washington (2005)
21. R.H. Bordini, J.F. Huebner, M. Wooldridge, *Programming Multi-Agent Systems in AgentSpeak using Jason* (Wiley, Hoboken, 2007)
22. V. Bulusu, R. Sengupta, V. Polishchuk, L. Sedov, Cooperative and non-cooperative UAS traffic volumes, in *2017 International Conference on Unmanned Aircraft Systems, ICUAS 2017, Miami* (2017)
23. P. Caillou, B. Gandou, A. Grignard, C.Q. Truoung, P. Taillandier, A simple to use BDI architecture for agent based modeling and simulation, in *The 11th Conference of the European Social Simulation Association, Groningen* (2015)
24. A. Cherniak, Explorng behavioral patterns in complex adaptive systems. Ph.D. Thesis, University of Pittsburgh (2014)
25. H. Choset, K.M. Lynch, S. Hutchinson, G. Kantor, W. Burgard, L.E. Kavraki, S. Thrun, *Principles of Robot Motion Theory, Algorithms, and Implementation* (MIT Press, Cambridge, 2005)
26. J. Christopher Beck, N. Wilson, Proactive algorithms for scheduling with probabilistic durations., in *IJCAI International Joint Conference on Artificial Intelligence*, vol. 28 (2005), pp. 1201–1206
27. M.P. Clay, N. Simanyi, Renyi's parking problem revisited. Stoch. Dyn. **16**, 1–12 (2014)
28. N. Hansen, Y. Akimoto, P. Baudis, CMA-ES/Pycma on Github (Zenodo, 2019). https://doi.org/10.5281/zenodo.2559634, February
29. R. D'Andrea, Can drones deliver? IEEE Trans. Autom. Sci. Eng. **11**(3), 647–648 (2014)
30. F. Darema, Dynamic data driven applications systems: a new paradigm for application simulations and measurements, in *Computational Science - ICCS 2004*, ed. by M. Bubak, G.D. van Albada, P.M.A. Sloot, J. Dongarra (Springer, Berlin 2004), pp. 662–669
31. L. De Raedt, A. Kimmig, H. Toivonen, ProbLog: a probabilistic prolog and its application in link discovery, in *International Joint Conference on Artificial Intelligence, Hyderabad* (Elsevier, Amsterdam, 2007)
32. E. Dietrich, A.B. Markman, *Cognitive Dynamics: Conceptual and Representational Change in Humans and Machines* (Psychology Press, New York, 2014)
33. P. Domingos, D. Lowd, *Markov Logic: An Interface Layer for Artificial Intelligence* (Morgan and Claypool, San Rafael, 2009)
34. J.S. Duncan, Instrument procedures handbook. Technical Report FAA-H-8083-16B, Federal Aviation Administration, Washington (2017)
35. D. Easley, J. Kleinberg, *Networks, Crowds, and Markets: Reasoning About a Highly Connected World* (Cambridge University Press, Cambridge, 2010)
36. J.A. Fax, R.M. Murray, Information flow and cooperative control of vehicle formations. IEEE Trans. Autom. Control **49**(9), 1465–1476 (2004)
37. Federal Aviation Administration, FAA aerospace forecast fiscal years 2020–2040. Technical report, Federal Aviation Administration (2020)
38. H.D. Friedman, D. Rothman, J.K. Mackenzie, Solution to: an unfriendly seating arrangement (Problem 62–3). SIAM Rev. **6**(2), 180–182 (1964)

39. M.L. Gargano, A. Weissenseel, J.F. Malerba, M. Lewinter, Discrete Renyi parking constants. Technical Report, Pace University, New York (2005)
40. M. Georgeff, B. Pell, M. Pollack, M. Tambe, M. Wooldridge, The belief-desire-intention model of agency, in *Intelligent Agents V: Agents Theories, Architectures, and Languages*, ed. by J.P. Müller, A.S. Rao, M.P. Singh (Springer, Berlin, 1999), pp. 1–10
41. German Federal Bureau of Aircraft Accidents Investigation, Investigation report. Technical report, Bundesstelle für Flugunfalluntersuchung, Braunschweig (2004)
42. DDER SPIEGEL. (2019, June 16). Deutsche erhalten mehr Pakete als andere Nationens. DER SPIEGEL, Hamburg, Germany. Retrieved January 6, 2020, from https://www.spiegel. de/wirtschaft/unternehmen/pakete-in-deutschland-wird-mehr-verschickt-als-in-ande-ren-nat-ionen-a-1272662.html
43. Global UTM Association, UAS traffic management architecture. Lausanne (2017)
44. V. Gogate, P. Domingos, Probabilistic theorem proving. Commun. ACM **59**(7), 107–115 (2016)
45. ddss/assets at master interuss/dss. (n.d.). GitHub. Retrieved June 1, 2020, from https://github. com/interuss/dss/tree/master/assets
46. T.C. Henderson, A. Mitiche, R. Simmons, X. Fan, A preliminary study of probabilistic argumentation. Technical Report UUCS-17-001, University of Utah (2017)
47. T.C. Henderson, R. Simmons, B. Serbinowski, M. Cline, D. Sacharny, X. Fan, A. Mitiche, Probabilistic sentence satisfiability: an approach to PSAT. Artif. Intell. **278**, 103199 (2020)
48. A. Hunter, A probabilistic approach to modelling uncertain logical arguments. Int. J. Approx. Reason. **54**(1), 47–81 (2013)
49. ICAO, Doc 9854 AN/458 - global air traffic management operational concept. International Civil Aviation Organization (2005), p. 82
50. Institute of Informatics, Belief, desire and intention agents. Web Page for Research Group (2019)
51. D.-S. Jang, C.A. Ippolito, S. Sankararaman, V. Stepanyan, Concepts of airspace structures and system analysis for UAS traffic flows for urban areas, in *AIAA Information Systems-AIAA Infotech @ Aerospace, Grapevine* (American Institute of Aeronautics and Astronautics, Reston, 2017)
52. M.R. Jardin, Analytical relationships between conflict counts and air-traffic density. J. Guid. Control Dyn. **28**(6), 1150–1156 (2005)
53. M. Johnson, J. Larrow, UAS traffic management conflict management model. Technical report, FAA-NASA UTM Research Transition Team: Sense and Avoid Working Group (2020)
54. A. Jones, L.C. Rabelo, A.T. Sharawi, Survey of job shop scheduling techniques, in *Wiley Encyclopedia of Electrical and Electronics Engineering* (Wiley, Hoboken, 1999)
55. J.B. Kenney, Dedicated short-range communications (DSRC) standards in the united states. Proc. IEEE **99**(7), 1162–1182 (2011)
56. M.J. Kochenderfer, C. Amato, G. Chowdhary, J.P. How, H.J.D. Reynolds, J.R. Thornton, P.A. Torres-Carrasquillo, N. Kemal Üre, J. Vian, *Decision Making Under Uncertainty: Theory and Application*, 1st edn. (The MIT Press, Cambridge, 2015)
57. P. Kopardekar, J. Rios, T. Prevot, M. Johnson, J. Jung, J.E. Robinson III, Unmanned aircraft system traffic management (UTM) concept of operations, in *16th AIAA Aviation Technology, Integration, and Operations Conference* (AIAA Aviation, Washington, 2016)
58. M. Kothari, I. Postlewaite, D.-W. Gu, UAV path following in windy urban environments. J. Intell. Robot. Syst. **74**, 1013–1028 (2014)
59. R. Kowalski, P.J. Hayes, Semantic trees in automatic theorem proving, in *Automation of Reasoning*, ed. by J.J. Siekmann, G. Wrightson, (Springer, Berlin, 1983), pp. 217–232
60. S.M. LaValle, *Planning Algorithms* (Cambridge University Press, Cambridge, 2006)
61. H. Li, N. Oren, T. Norman, Probabilistic argumentation frameworks, in *Proceedings of 1st International Workshop on the Theory and Applications of Formal Argumentation, Beijing* (2011)
62. A.J. Lohn, What's the buzz? The city-scale impact of drone delivery. Technical Report RR1718, RAND Corporation, Santa Monica (2017)

63. L. Martin, C. Wolter, K. Jobe, M. Manzano, S. Blandin, M. Cencetti, L. Claudatos, J. Mercer, J. Homola, TCL4 UTM (UAS traffic management) Nevada 2019 flight tests, airspace operations laboratory (AOL) report. Technical Report, National Aeronautics and Space Administration (2020)

64. K. Nagel, S. Rasmussen, Traffic at the edge of chaos, in *Artificial Life IV: Proceedings of the Fourth International Workshop on the Synthesis and Simulation of Living Systems* (MIT Press, Cambridge, 1994), pp. 222–235

65. A Namatame, S.-H. Chen, *Agent Based Modeling and Network Dynamics* (Oxford University Press, Oxford, 2016)

66. NASA, Unmanned aircraft systems (UAS) traffic management (UTM) concept of operations, V2.0. Technical Report, Federal Aviation Administration, Washington (2020)

67. National Aeronautics and Space Administration (NASA), Urban air mobility market study executive summary. Technical Report, National Aeronautics and Space Administration (NASA) (2018)

68. National Research Council (U.S.), Transportation Research Board, Highway capacity manual. Transportation Research Board, National Research Council, Washington (2000)

69. G.L. Neuhauser, A.H.G. Rinnooy Kan, M.J. Todd, *Optimization* (North-Holland, New York, 1989)

70. G.F. Newell, A simplified car-following theory: a lower order model. Transp. Res. B Methodol. **36**(3), 195–205 (2002)

71. N. Nilsson, Probabilistic logic. Artif. Intell. J. **28**, 71–87 (1986)

72. G. Oriz-Hernandez, J.F. Huebner, R.H. Bordini, A. Guerra-Hernandez, G.J. Hoyos-Rivera, N. Cruz-Ramirez, A namespace approach for modularity in BDI programming languages, in *Engineering Multi-Agent Systems: 4th International Workshop*, ed. by M. Baldoni, J.P. Mueller, I. Nunes, R. Zalila-Wenkstern (Springer, Singapore, 2016), pp. 117–135

73. C.H. Papadimitriou, K. Steiglitz, *Combinatorial Optimization: Algorithms and Complexity* (Dover Publications, Mineola, 1988)

74. M. Patriksson, *The Traffic Assignment Problem: Models and Methods* (Dover Publications, Mineola, 2015)

75. M.L. Pinedo, *Scheduling: Theory, Algorithms, and Systems*, 5th edn. (Springer, New York, 2016)

76. S.F. Railsback, V. Grimm, *Agent-Based and Individual-Based Modeling* (Princeton University Press, Princeton, 2012)

77. A.S. Rao, AgentSpeack(L): BDI agents speak out in a logical computable language, in *Proceedings of the 7th European Workshop on Modeling Autonomous Agents in a Multi-Agent World*, ed. by W. Van de Velde, J.W. Perram (Springer, Eindhoven, 1996), pp. 42–55

78. C. Reiche, R. Goyal, A. Cohen, J. Serrao, S. Kimmel, C. Fernando, S. Shaheen, Urban air mobility market study. Technical report, National Aeronautics and Space Administration (NASA) (2018)

79. G. Rens, D. Moodley, A hybrid POMDP-BDI agent architecture with online stochastic planning and plan caching. Cognit. Syst. Res. **43**, 1–20 (2017)

80. A. Rényi, On a one-dimensional problem concerning random space filling. Publ. Math. Inst. Hungarian Acad. Sci. **3**, 109–127 (1958)

81. M. Richardson, P. Domingos, Markov logic networks. Mach. Learn. **62**(1–2), 107–136 (2006)

82. J.L. Rios, UAS traffic management (UTM) project strategic deconfliction: system requirements final report. Technical Report NASA Report, NASA, Moffet Field (2018)

83. J. Rios, J. Lohn, A comparison of optimization approaches for nationwide traffic flow management, in *AIAA Guidance, Navigation, and Control Conference and Exhibit* (American Institute of Aeronautics and Astronautics, Reston, 2009)

84. J. Rios, K. Ross, Delay optimization for airspace capacity management with runtime and equity considerations, in *AIAA Guidance, Navigation and Control Conference and Exhibit* (2007), p. 10

85. J.L. Rios, I.S. Smith, P. Venkatesan, D.R. Smith, V. Baskaran, S. Jorcak, S. Iyer, P. Verma, UTM UAS service supplier development. Technical Report NASA/TM-2018-220050, NASA (2018)

86. S. Russell, P. Norvig, *Artificial Intelligence: A Modern Approach* (Upper Saddle River, Prentice-Hall, 2001)

87. S. Russell, D. Dewey, M. Tegmark, Research priorities for robust and beneficial artificial intelligence. AI Mag. **36**(4), 105–114 (2015)

88. D. Sacharny, T.C. Henderson, Optimal policies in complex large-scale UAS traffic management, in *IEEE Conference on Industrial Cyber-Physical Systems, Taipei* (2019)

89. D. Sacharny, T.C. Henderson, A lane-based approach for large-scale strategic conflict management for UAS service suppliers, in *2019 International Conference on Unmanned Aircraft Systems (ICUAS)* 2019 (Institute of Electrical and Electronics Engineers (IEEE), Atlanta, 2019), pp. 937–945

90. D. Sacharny, T.C. Henderson, A lane-based approach for large-scale strategic conflict management for UAS service suppliers, in *IEEE International Conference on Unmanned Aerial Systems, Atlanta* (2019)

91. D. Sacharny, T.C. Henderson, Optimal policies in complex large-scale UAS traffic management, in *IEEE Conference on Industrial Cyber-Physical Systems, Taipei* (2019)

92. D. Sacharny, T.C. Henderson, Optimal policies in complex large-scale UAS traffic management, in *IEEE International Conference on Industrial Cyber-Physical Systems, Taipei* (2019)

93. D. Sacharny, C. Liu, Strategic deployment of drone centers and fleet size planning for drone delivery, Report No. UT-21.33. Technical Report, Utah Department of Transportation, Salt Lake City (2021)

94. D. Sacharny, T.C. Henderson, A. Mitiche, R. Simmons, T. Welker, X. Fan, Breccia: unified probabilisitic dynamic geospatial intelligence, in *IEEE Conference on Intelligent Robots and Systems (IROS 2017 Late Breaking Paper), Vancouver* (2017)

95. D. Sacharny, T.C. Henderson, A. Mitiche, R. Simmons, T. Welker, X. Fan, BRECCIA: a multi-agent data fusion and decision support system for dynamic mission planning, in *2nd Conference on Dynamic Data Driven Application Systems (DDDAS 2017), Cambridge* (2017)

96. D. Sacharny, T.C. Henderson, R. Simmons, A. Mitiche, T. Welker, X. Fan, BRECCIA: a novel multi-source fusion framework for dynamic geospatial data analysis, in *IEEE Conference on Multisensor Fusion and Integration, Daegu* (2017)

97. D. Sacharny, T. Henderson, M. Cline, B. Russon, Reinforcement learning at the cognitive level in a belief, desire, intention UAS agent. Technical Report UUCS-20-013, University of Utah (2020)

98. D. Sacharny, T.C. Henderson, M. Cline, Large-scale UAS traffic management (UTM) structure, in *IEEE Multisensor Fusion and Integration Conference, Karlsruhe* (2020)

99. D. Sacharny, T.C. Henderson, M. Cline, B. Russon, E. Guo, FAA-NASA vs. lane-based strategic deconfliction, in *IEEE Multisensor Fusion and Integration Conference, Karlsruhe* (2020)

100. D. Sacharny, T.C. Henderson, E. Guo, A DDDAS protocol for real-time UAS flight coordination, in *InfoSymbiotics/Dynamic Data Driven Applications Systems Conference, Boston* (2020)

101. D. Sacharny, T.C. Henderson, M. Cline, B. Russon, Reinforcement learning at the cognitive level in a belief, desire, intention UAS agent, in *Intelligent Autonomous Systems Conference, Singapore* (2021), pp. 431–442

102. M. Saha, P. Isto, Multi-robot motion planning by incremental coordination, in *IEEE International Conference on Intelligent Robots and Systems* (2006), pp. 5960–5963

103. A. Schadschneider, M. Schreckenberg, Cellular automation models and traffic flow. J. Phys. A: Math. Gen. **26**(15), L679–L683 (1993)

104. M. Schut, M. Wooldridge, Principles of intention reconsideration, in *Proceedings of the Fifth International Conference on Autonomous Agents, AGENTS'01* (Association for Computing Machinery, New York, 2001), pp. 340–347

105. M. Schut, M. Wooldridge, S. Parsons, On partially observable MDPs and BDI models, in *Foundations and Applications of Multi-Agent Systems*, ed. by M. d'Inverno, M. Luck, M. Fisher, C. Preist (Springer, Berlin, 2002), pp. 243–259

106. L. Sedov, V. Polishchuk, Centralized and distributed UTM in layered airspace, in *8th International Conference on Research in Air Transportation* (2018), pp. 1–8
107. G.I. Simari, S. Parsons, On the relationship between MDPs and the BDI architecture, in *Proceedings of the Fifth International Joint Conference on Autonomous Agents and Multiagent Systems, AAMAS'06* (Association for Computing Machinery, New York, 2006), pp. 1041–1048
108. E. Sunil, J. Hoekstra, J. Ellerbroek, F. Bussink, D. Nieuwenhuisen, A. Vidosavljevic, S. Kern, Metropolis: relating airspace structure and capacity for extreme traffic densities, in *11th USA/Europe Air Traffic Management Research and Development Seminar, Lisbon* (2015)
109. L.D. Swartzentruber, Improving path planning of unmanned aerial vehicles in an immersive environment using metapaths and terrain information. Ph.D. Thesis, Iowa State University (2009)
110. Texas A&M Transportation Institute, Traffic management centers. https://static.tti.tamu.edu/tti.tamu.edu/documents/policy/congestion-mitigation/traffic-management-centers.pdf. Accessed Dec. 2021
111. M. Thimm, A probabilistic semantics for abstract argumentation, in *Proceedings of 20th European Conference on Artificial Intelligence, Montpellier* (2012)
112. D.P. Thipphavong, R.D. Apaza, B.E. Barmore, V. Battiste, B.K. Burian, Q.V. Dao, M.S. Feary, S. Go, K.H. Goodrich, J.R. Homola, H.R. Idris, P.H. Kopardekar, J.B. Lachter, N.A. Neogi, H.K. Ng, R.M. Oseguera-Lohr, M.D. Patterson, S.A. Verma, Urban air mobility airspace integration concepts and considerations, in *2018 Aviation Technology, Integration and Operations Conference* (AIAA, Atlanta, 2018)
113. M. Tra, E. Sunil, J. Ellerbroek, J. Hoekstra, Modeling the intrinsic safety of unstructured and layered airspace designs, in *12th USA/Europe Air Traffic Management Research and Development Seminar, Seattle* (2017)
114. M. Treiber, A. Kesting, *Traffic Flow Dynamics* (Springer, Berlin, 2013)
115. Utah Geospatial Resource Center (UGRC), Building footprints (2021). https://gis.utah.gov/data/location/building-footprint
116. Utah Geospatial Resource Center (UGRC), Projected populations and households for TAZs (2021). https://gis.utah.gov/data/demographic/projections/#populationhouseholdstaz
117. Utah Geospatial Resource Center (UGRC), Utah roads (2021). https://gis.utah.gov/data/transportation/roads-system/
118. Utah Geospatial Resource Center (UGRC), Zip code PO boxes (2021). https://gis.utah.gov/data/location/u-s-postal-service/#ZipCodePOBoxes
119. J. van den Berg, J. Snoeyink, M. Lin, D. Manocha, Centralized path planning for multiple robots: optimal decoupling into sequential plans, in *Robotics: Science and Systems V* (2009)
120. A.S. Vezhnevets, S. Osindero, T. Schaul, N. Heess, M. Jaderberg, D. Silver, K. Kavukcuoglu, FeUdal networks for hierarchical reinforcement learning, in *Proceedings of the 34th International Conference on Machine Learning - Volume 70, Sydney*. JMLR.org (2017), pp. 3540–3549
121. I. Zelinka, V. Snasel, A. Abraham, *Handbook of Optimization* (Springer, Berlin, 2013)

Index

© The Author(s), under exclusive license to Springer Nature Switzerland AG 2022   191
D. Sacharny, T. C. Henderson, *Lane-Based Unmanned Aircraft Systems Traffic
Management*, Unmanned System Technologies,
https://doi.org/10.1007/978-3-030-98574-5

Printed in the United States
by Baker & Taylor Publisher Services